GETTING TO KNOW

ArcGIS

Enterprise

GETTING TO KNOW

ArcGIS®
Enterprise

Jon Emch
Diana Muresan
Travis Ormsby

Esri Press
REDLANDS | CALIFORNIA

Esri Press, 380 New York Street, Redlands, California 92373-8100

Copyright © 2025 Esri

All rights reserved.

Printed in the United States of America

29 28 27 26 25   1 2 3 4 5 6 7 8 9 10

Library of Congress Control Number: 2025942560

ISBN: 9781589487932

Contents

Acknowledgments

This book represents the perspectives and experiences of some of the brightest ArcGIS Enterprise practitioners in the world. We want to thank the members of the ArcGIS Enterprise community within Esri for supporting us in scoping out what would become *Getting to Know ArcGIS Enterprise.*

Special thanks to our technical editors, who verified each chapter for technical accuracy: Jonathan Murphy, Mark Carlson, Jill Edstrom-Shoemaker, Randall Williams, Sam Libby, Maggie Busek, Marley Geddes, Kaitlyn Stevens, Jessica Mignault, Nana Dei, Sarah Hanson, Noah Krach, Faris Bouran, Andrew Sakowicz, Christopher Pawlyszyn, and Anne Reuland.

We also thank Claudia Naber, Maryam Mafuri, Mark Henry, Monica McGregor, and Carolyn Schatz at Esri Press for their unwavering support and assistance in developing the work.

A special thank-you to Christian Wells, Rebecca Richman, and Philip Heede, who supported the early days of this work. Your wisdom and insight helped us make this journey successful and complete.

Additional thanks to the analysts, tech and readiness leads, and managers of Esri Support Services, whose knowledge and experiences aided in the creation of the scenarios of this book.

To our families and friends, we thank you for your encouragement and support. Writing a book turned out to be a tougher undertaking than we first imagined, and we could not have finished without your support.

Finally, we thank the worldwide GIS community for doing amazing work with ArcGIS technology. Your stories and ideas inspire all of us to create a more perfect product and to author works that support your mission.

Introduction

"Hello, I am a GIS analyst from company XYZ, and my ArcGIS® Enterprise administrator has recently left the organization. I've been given an administrator account, but I'm unfamiliar with the different functions and responsibilities I must do to keep the deployment running. What is ArcGIS Enterprise?"

The authors of this book have answered this question countless times over their careers, whether at user conferences, over technical support calls, or teaching instructor-led courses. It's a common scenario for the way GIS analysts manage an ArcGIS Enterprise environment. Although analysts may be familiar with the role of ArcGIS Enterprise in various workflows, the background administrative knowledge may not be readily available to them.

Getting to Know ArcGIS Enterprise was written to help new ArcGIS Enterprise administrators succeed with administrative, technical, and practical workflows. Although not exhaustive, this book is designed to serve as an effective base of understanding to help ArcGIS Enterprise administrators build their knowledge. It's also a handy resource for seasoned administrators who want a quick reference to other parts of the software. The authoring team gauged each topic to be appropriate for most new and intermediate ArcGIS Enterprise administrators.

About the authors

Who is the team? We are three practitioners of ArcGIS Enterprise within Esri®. In assembling the book, our goal was to capture a variety of voices and perspectives. To achieve this, the coauthors came together from different backgrounds and perspectives to create this volume.

Jon Emch is a senior technical lead for Esri Technical Support. Jon, who started working with ArcGIS Enterprise as an intern in 2017, seeks to promote technical literacy at all levels within Esri. He has written several blogs on topics ranging from distributed collaborations to patching. He leads a team of technical leads that specializes in understanding and solving the most complex ArcGIS Enterprise support cases. Jon received a BA in geography from Virginia Polytechnic Institute and State University in Blacksburg, Virginia.

Diana Muresan is a senior product engineer on the geodatabase software development team. In her role, Diana applies GIS to real-world workflows in guided tutorials, demos, stories with interactive maps, and blog posts to help users maximize the potential of spatial data management. Diana received a BA and MS in geomatics from Babeş-Bolyai University in Romania and an MEng in civil engineering and GIS from the University of Colorado, Denver.

Travis Ormsby is a senior product engineer on the ArcGIS Enterprise documentation team. His eclectic career has taken him from flint knapping stone tools to deploying cloud-native applications on Kubernetes. Travis, an educator by trade, helps people learn to use ArcGIS effectively through instructor-led training, web course development, blog posts, videos, and official help documentation. Travis received a BA in anthropology from Grinnell College in Iowa, an MA in teaching from the University of St. Thomas in Minnesota, and an MGIS from the University of Minnesota.

In addition to the authors, this work was reviewed by many subject matter experts from Esri, who added insights to the work and helped ensure the accuracy of the technical details.

About the book

As you read *Getting to Know ArcGIS Enterprise*, remember that ArcGIS Enterprise is complimentary to ArcGIS Pro and ArcGIS Online. Many workflows and concepts in this book work in tandem with both of those products, and many more. Consider reviewing *Top 20 Essential Skills for ArcGIS Online* and *GIS Tutorial for ArcGIS Pro 3.4* from Esri Press to supplement your knowledge.

The book is divided into four sections, each covering a logical part of using ArcGIS Enterprise:

1. **Implementing ArcGIS Enterprise**: This section covers the basics of what is required for a functional ArcGIS Enterprise base deployment. These basics include understanding prerequisites, the anatomy of base deployment, installing and configuring ArcGIS Enterprise, securing ArcGIS Enterprise, and extending the base deployment.

2. **Administering ArcGIS Enterprise**: This section describes how to properly administer ArcGIS Enterprise deployment. We will introduce a way for managing content, users, and groups. Additionally, this section covers how to manage content from different data sources.

3. **Using Common Workflows in ArcGIS Enterprise**: This section is a primer on fundamental workflows that support a wide range of feature service usage. Some workflows covered include how to publish from a geodatabase, consuming web layers for editing and spatial analysis, and how to share your work.

4. **Maintaining and Troubleshooting in ArcGIS Enterprise**: This section is a collection of topics that deal with running ArcGIS Enterprise in a safe configuration. Topics covered in this section include how to back up and restore ArcGIS Enterprise, how to find and apply patches, and how to prepare for a system upgrade.

Special font styles used

In this book, student input for a work step is designated by **colored highlighted text**. The computer user interface elements are styled in **bold font**.

Fictional user stories

Throughout this book, we've included several fictional user stories to articulate the lessons of each chapter. You will find some background information on each of these fictional entities. Throughout this book, we've included many fictional stories that illustrate the types of issues we've worked on with ArcGIS Enterprise users. Those issues might be more or less relevant to different types of organizations, so we've invented three different fictional entities that represent the broad range of organizations that use ArcGIS Enterprise. Here is a some background information on these fictional entities to keep in mind as you read their stories in this book:

- **SuperBiz International**: A large logistics company of 15,000 employees, SuperBiz is a global leader in delivering multimodal transportation solutions for its customers. It has a wide variety of needs related to spatial data, such as tracking its vehicle fleets in real time and performing sophisticated spatial analyses. SuperBiz also must securely share some of its data with its external partners.
- **Medio County**: With a current population of 100,000 people, Medio County is moderately sized but growing rapidly. The county maintains spatial datasets that need to be publicly available, such as parcel data, and others that need to remain private, such as the addresses of emergency service responses. The county also provides GIS services for several small municipalities in the county that do not have their own GIS capabilities. The county's ArcGIS Enterprise deployment is complex enough that it needs to deploy it on multiple machines running on Amazon Web Services (AWS).
- **Becken Pond Conservation Society**: This small nonprofit organization formed to rehabilitate a local wetland. The work of the society is supported by volunteers who collect samples and control invasive species. The society also has a sophisticated fundraising operation that pays for the construction of earthworks to restore the natural topography of the pond. A part-time GIS staffer maintains all the society's spatial datasets and integrates the data with its donor management system.

We are excited for you to start this book, which was written so that you can read the chapters in any order, depending on your interest. We hope that whether you are new to ArcGIS Enterprise or a more seasoned administrator, you will find this book a useful guide in your own journey with GIS.

PART 1
Implementing ArcGIS Enterprise

THE FIRST FIVE CHAPTERS OF THIS BOOK INTRODUCE THE CORE SET of decisions administrators should make before deciding how to best implement their ArcGIS Enterprise deployment. This section begins by reviewing prerequisites for the operating system, hardware, and security. It then introduces the ArcGIS Enterprise base deployment, which is made up of four components: Portal for ArcGIS, ArcGIS Server, ArcGIS Data Store, and two instances of ArcGIS Web Adaptor. Part 1 details important installation and security considerations for ArcGIS Enterprise and concludes with a section on extending the base deployment. Although each chapter in part 1 offers best practices, they are not in any way prescriptive; you may need to modify many of the steps introduced here to meet your organization's needs. These chapters will help you understand what factors need to be considered to ensure the successful deployment and long life of ArcGIS Enterprise for your organization.

CHAPTER 1
Understanding the prerequisites

Objectives
- Review system architecture requirements for ArcGIS Enterprise.
- Discuss operating system (OS) compatibility.
- Understand system resource considerations.
- Consider security before implementation.

Introduction
In this chapter, you'll discover the meaning of various prerequisites that should be considered before designing and deploying ArcGIS Enterprise. We will consider how different OSs come into play with ArcGIS Enterprise, as well as what self-hosted options are available and what factors should be considered when choosing your base infrastructure.

ArcGIS Enterprise is a part of the ArcGIS geospatial platform. The ArcGIS geospatial platform frames all ArcGIS software offerings as part of a greater whole, created to meet a wide variety of needs across the GIS industry. In the scope of this book, you will see references to additional software, such as ArcGIS Online, ArcGIS Monitor, and other applications. These pieces of software exist to add additional functionality to your organization by working with ArcGIS Enterprise in a variety of architecture constraints.

As a note, this chapter goes over basic information and considerations before ArcGIS Enterprise is deployed. Being an enterprise-grade software, it has many permutations and advanced deployment options that may be implemented in system design. We will acknowledge and discuss the benefits of pursuing more advanced system design later in this chapter.

ArcGIS Enterprise system basics 101

To someone just beginning to use ArcGIS Enterprise, let's start by explaining what enterprise software is. At its core, any enterprise software seeks to be scalable to the needs of an organization. Unlike desktop-based software, enterprise software seeks to fulfill the needs of many concurrent users. Each implemented feature asserts how to best fulfill the needs of an ever-growing organization. ArcGIS Enterprise follows industry best practices to be a scalable and secure technology. Consider the six architectural pillars of enterprise software:

- **Automation**: Enterprise software enables the use of scripts, webhooks, and other automation tools to streamline workflows.
- **Integration**: Enterprise software seeks to connect datasets and workflows from a wide variety of sources.
- **Observability**: Enterprise software allows administrators to see how the system is operating, catch potential issues proactively, and effectively respond to problems.
- **Performance and scalability**: Enterprise-level software can accommodate increases in the number of users or the size of the data footprint.
- **Reliability**: Enterprise software considers common failure conditions. High availability, backup strategies, recovery options, and workload separation fit into this mold.
- **Security**: Enterprise software implements security best practices to authenticate and authorize requests.

ArcGIS Enterprise aligns with these principles to fit a variety of user needs. At times, the number of options may seem overwhelming. When considering what type of deployment to pursue, it can be helpful to think of ArcGIS Enterprise as a *system*. To help illustrate what this system may look like, ask yourself the following questions:

- What are the primary needs for this deployment?
- How many users will this deployment need to serve?
- What kinds of data sources will I need to consume and account for?
- What are the primary workflows that my organization needs to address?

The answers to these questions are the beginning of an ArcGIS Enterprise implementation that is not only technically viable but also considers the core needs of an organization. Identifying these requirements early in the system design stage of an ArcGIS Enterprise deployment will make choosing prerequisites such as OSs and underlying architecture providers easier.

> *Note: Throughout this book, we will be making numerous references to the ArcGIS Well-Architected Framework and the ArcGIS Architecture Center (links.esri.com/GTKEnterprise-ArchCenter). This resource was created to bridge the IT-GIS knowledge gap and addresses key knowledge on understanding how to configure and create a resilient deployment that meets various needs.*

1

ArcGIS Enterprise architecture

Now that we've presented the basics of enterprise software, let's bring it back to ArcGIS Enterprise. ArcGIS Enterprise supports many popular self-hosted options. ArcGIS Enterprise runs on infrastructure that you or a trusted vendor control. This gives you maximum flexibility when configuring your architecture. It also means that you have responsibility for configuring ArcGIS Enterprise to meet your organization's needs. You can choose from several infrastructure options.

Some organizations want to deploy ArcGIS Enterprise on physical machines running in their own data centers. This option is the traditional model of server architecture, and many organizations have deep expertise in running their own physical infrastructure. It is a good option for organizations that need to keep their deployment disconnected from the internet. Building on this option, customers may also deploy ArcGIS Enterprise on virtualized environments that are disconnected from the internet.

The growing popularity of cloud-based resources and virtual machines has had a dramatic impact on the GIS software industry. Many organizations are taking a hybrid approach to their IT infrastructure. A 2023 poll from Zippia reported that 94 percent of organizations use some level of cloud services for their needs (links.esri.com/GTKEnterprise-ZippiaPoll). ArcGIS Enterprise was developed to support not only on-premises deployments such as traditional server architecture but also virtual machines and cloud offerings. This support can include running ArcGIS Enterprise on Linux and Microsoft Windows or a cloud native offering, such as ArcGIS Enterprise on Kubernetes®.

ArcGIS Enterprise also provides workload separation. Each major component of ArcGIS Enterprise can be placed on individual machines to separate the workload. We will discuss these components in chapter 2. More information on workload separation, as well as some strategies, can be found in the ArcGIS Architecture Center (links.esri.com/GTKEnterprise-WorkloadSeparation).

The operating system

So far, we've talked generally about how ArcGIS Enterprise works. For the rest of the chapter, we will cover specific prerequisites that contribute heavily to a functioning ArcGIS Enterprise deployment. Deciding on which OS to use requires a level of coordination with the rest of your organization. Although not the flashiest consideration, choosing the OS will have substantial downstream effects in the form of OS maintenance, updating the OS, and meeting organization-level standards.

ArcGIS Enterprise as a software set is built to accommodate the most popular OS being used for enterprise-level software. ArcGIS Enterprise supports Windows and a set of enterprise-level Linux options.

> Note: ArcGIS Enterprise recently added support for deployment on Kubernetes. Kubernetes is an open-source system for automated deployment, scaling, and management of containerized applications. For more details on this deployment option, refer to chapter 5.

When you decide on an OS, keep the following in mind:
- What OS is my organization mainly using for its architecture?
- Is my OS supported by the OS provider?

Let's take a closer look at what each of these questions implies.

What OS is my organization using?

The most important question to be answered, and the one that has the most weight in the decision, is what OS is your organization already familiar with? In most cases, it is best practice to keep ArcGIS Enterprise deployed on one type of OS, although some use cases may require deploying a mixture of OSs. It is best to avoid added complexity by deploying a Windows version of ArcGIS Enterprise within an organization that is primarily using Linux. As an ArcGIS Enterprise administrator, you should capitalize on the knowledge and skillset of your colleagues in IT to the best of your ability. To this point, it's best to stay consistent with the most familiar OS being used by your organization.

Is my OS supported by the OS provider?

Also understand the life cycle of the OS before deploying ArcGIS Enterprise. By planning for an OS version that won't retire anytime soon, you can reduce the architectural complexities you need to consider. Researching the life cycle of the underlying OS and understanding the implications of using that version are key to building a resilient system.

Although no functional differences exist between Linux and Windows versions of ArcGIS Enterprise, the installation experience can vary between these versions. For example, ArcGIS Web Adaptor is a component that relies on either Microsoft's Internet Information System (IIS) when running on Windows or a Java component when running on Windows or Linux. Although the function is the same, the underlying configuration and installation vary greatly. Keep this in mind when you deploy ArcGIS Enterprise.

The hardware

One of the most common questions that new users of ArcGIS Enterprise have is about hardware considerations. Be sure that your environment has enough resources not only to run ArcGIS Enterprise effectively but also to run the OS and any other applications that may be present on the system.

How much is enough? How do I decide how many resources I need? This may seem like a simple question to answer, but it is multifaceted and depends on how you are deploying the system and how distributed it is.

ArcGIS Enterprise documentation contains details of the minimum requirements that each component of ArcGIS Enterprise needs to function properly. Note that these minimums are the least number of resources required to install, run, and complete basic functions on a system. You must therefore consider any additional usage when you decide on the initial resource allocation of ArcGIS Enterprise.

Deciding how to allocate resources is not a "one size fits all" approach. Many variables play into what may be appropriate for each user. Here are some variables to consider when deciding on resources:

- Ensure that system minimums are enough to install and operate the software.
- Consider how many users are going to need to access the deployment.
- Because ArcGIS Enterprise supports hosted and referenced data, understand the database footprint and what workflows may be run against these data sources.

Note: The Architecture Center includes a list of case studies that describe the types of ArcGIS Enterprise deployments for various use cases and industries. These case studies delve into resource management and include architecture diagrams alongside resource allocations for each component. Read more about these test studies in the Test Study Gallery in the ArcGIS Architecture Center (links.esri.com/GTKEnterprise-TestStudies).

The key takeaway is to remain flexible. It is common for newer implementations to change the hardware resources available for ArcGIS Enterprise once load figures become available. Additionally, ArcGIS Enterprise can be extended with different ArcGIS Server roles depending on customer and system needs. These can be configured once a base deployment is achieved as those needs arise. It is wise to set up resource monitoring for ArcGIS Enterprise to keep track of how well your deployment is functioning.

Note: Dynamic scaling for unexpected load can be achieved through most ArcGIS Enterprise implementations. However, ArcGIS Enterprise on Kubernetes was designed with containers in mind, enabling dynamic scaling of containers and services as needed. To read more about this, refer to the documentation on service scaling (links.esri.com/GTKEnterprise-ServiceScaling).

Security considerations

ArcGIS Enterprise is designed to be deployed in a variety of network configurations. If legal obligations prevent data access outside of an internal network or other security requirements, ArcGIS Enterprise will function as a cohesive system without an external network connection in what's called a disconnected environment. Alternatively, ArcGIS Enterprise can be deployed to be accessible from the internet through a security system and the use of a perimeter network.

A common implementation decision at this level is to adopt a hybrid system. This can be done in a variety of ways, but one of the most effective is to add arcgis.com domains to the *allow list* in the firewall. You might want to do this to take advantage of various functionalities in ArcGIS Online that are available for use in ArcGIS Enterprise. For example:

- Enable the use of online basemaps to give users different information layers when making maps and apps.
- Augment user content through Esri-curated data through ArcGIS Living Atlas of the World.
- Enable easier collaboration between users outside and inside your network through distributed collaborations.

Each of these features has different requirements, which can be found in the documentation at links.esri.com/GTKEnterprise-AGOLCollab.

Maintenance

Once ArcGIS Enterprise is deployed, ArcGIS Enterprise administrators need to work with their organization's IT department to maintain the software environment. Active maintenance includes patching ArcGIS Enterprise and other necessary software, upgrading ArcGIS Enterprise, adding system resources when required, and creating backups of the environment. We will explore maintenance tasks in greater detail in part 4.

Fictional user story

The Becken Pond Conservation Society recently received a grant that will enable them to implement geospatial capabilities strategically instead of relying on their current ad hoc system. Jim Yazzie, the GIS administrator, first catalogs the society's spatial data needs.

Volunteers document the location of invasive species and collect soil and water samples around the pond, where they have poor internet access.

Field data must go through an advanced data validation process for quality assurance.

Spatial data must be integrated with the society's donor management system to enable targeted fund-raising based on donor profiles.

Jim knows the ArcGIS Architecture Center provides detailed information about the system patterns. He reads the overviews to match the society's needs with the appropriate system pattern. Through this process, Jim determines that the mobile operations and offline data management system pattern will enable all the capabilities the society requires.

After identifying the appropriate system pattern, Jim starts to investigate the appropriate deployment pattern for that system. Using the ArcGIS Architecture Center as a guide, he compares deployment patterns.

A PaaS deployment using ArcGIS Location Platform does not support mobile operations. Neither ArcGIS Enterprise on Kubernetes nor a SaaS deployment on ArcGIS Online support the necessary advanced data validation capabilities. But ArcGIS Enterprise on Windows and Linux support all the capabilities the society needs.

Investigating the base architecture of deploying this system pattern using ArcGIS Enterprise on Windows and Linux, he sees this architecture diagram:

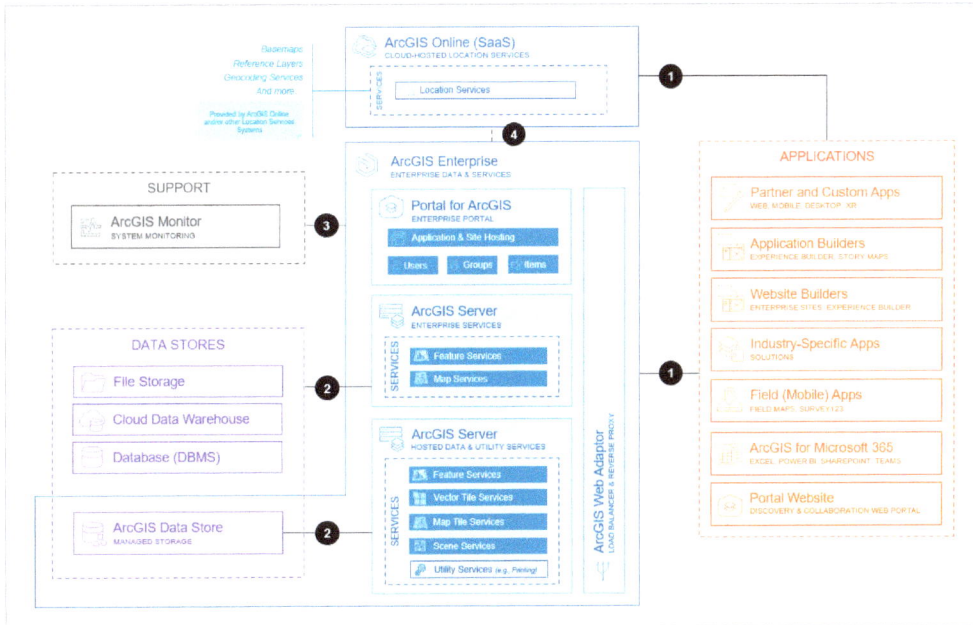

Figure 1.1. Sample base deployment of ArcGIS Enterprise on Windows or Linux.

Through his investigation, Jim does not believe ArcGIS Online integration will be necessary because the society has its own basemaps and reference layers and does not require geocoding or routing services. The deployment will also host a relatively small number of services, so it is unnecessary to separate the workload among multiple ArcGIS Server sites. To monitor the deployment, Jim uses ArcGIS Server statistics and logs. Jim also works with his existing GIS analysts to identify a referenced geodatabase that must be used with the deployment. Having defined these requirements, Jim draws up a proposed architecture diagram.

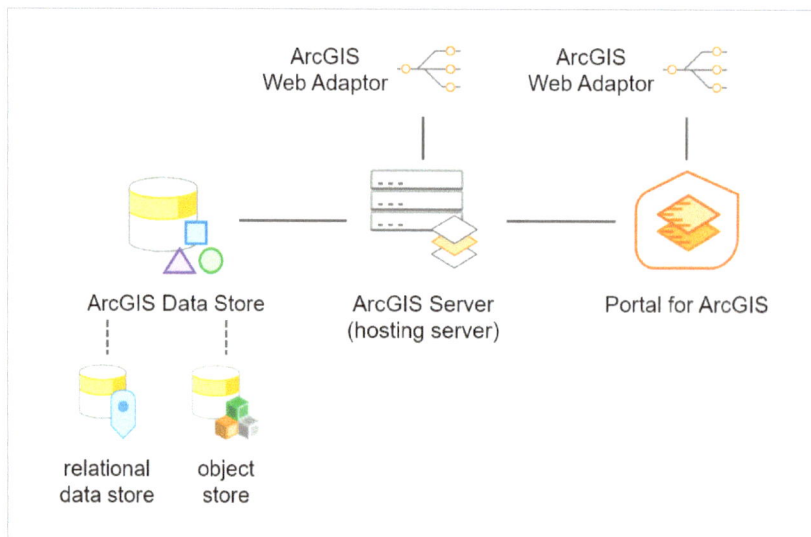

Figure 1.2. Basic architecture of ArcGIS Enterprise on Windows and Linux supports society's needs.

As Jim creates the architecture diagram, he begins to work with his IT resources on scoping out the necessary architecture to support the deployment. Considering that the Brecken society uses Windows in the cloud, he decides to work with existing architecture patterns. As the minimum viable deployment becomes clear, the IT team works with the administrator on creating a maintenance plan and schedule for updates and backups.

These steps have prepared Jim, the IT department, and the Brecken Pond Conservation Society to deploy their ArcGIS Enterprise environment successfully. This approach also ensured that key people and resources were identified and brought into the architecture side of the system before it was created.

1

Tutorial 1: Examine system patterns

1. In your browser, navigate to the ArcGIS Architecture Center at **architecture.arcgis.com**.

2. At the top of the page, click the **System Patterns** tab.

3. Review the ArcGIS system patterns.
   - Which system pattern or patterns supports the capabilities your organization needs?

4. On the left of the page, expand one of the system patterns that supports your organization's needs and read the overview.

5. Under the **Location services** system pattern, read the section on **Selecting a deployment pattern** to compare the capabilities of each deployment pattern.
   - Based on your organization's needs, which deployment pattern or patterns fully support those needs?

6. Expand the **Deployment patterns** section and click **Windows and Linux**.
   - What are the key components of a base architecture for this system pattern in Windows and Linux?
   - Which components, if any, of this base architecture are unnecessary for your organization's needs?
   - Does this architecture provide all the capabilities your organization needs? If not, which other system patterns will you need to implement to provide those capabilities?

Summary

In this chapter, we discussed ArcGIS Enterprise as an enterprise-grade system. Esri identified the pillars of enterprise system software: automation, integration, observability, performance and scalability, reliability, and security. Throughout the rest of the book, we will refer to these pillars to articulate how ArcGIS Enterprise fulfills them.

CHAPTER 2

Understanding the ArcGIS Enterprise base deployment

Objectives

- Explore the history of ArcGIS Enterprise.
- Define the ArcGIS Enterprise base deployment.
- Understand each base component of ArcGIS Enterprise.

Introduction

ArcGIS Enterprise is the foundational software system for GIS, powering mapping and visualization, analytics, and data management. This definition of the software is the most succinct way to summarize a software solution that can scale to accommodate organizations that have thousands of users while still providing this base level of service. Our goal in this chapter is to distill this definition into its key parts to promote your understanding.

Once you finish this chapter, you will have a fundamental understanding of ArcGIS Enterprise. We will begin by defining ArcGIS Enterprise base deployment. From there, we will explore the components that make up the base deployment. We will explain what each of the components does in the system, focusing on their role to a user within the organization.

At the end of this chapter, you will apply what you've learned by examining how such a deployment can scale to meet the ever-growing demands of an organization.

ArcGIS Enterprise: A brief history

To best understand why ArcGIS Enterprise is architected the way it is, let's start with a brief history lesson. Before ArcGIS Enterprise and modern GIS, or even web GIS, the technology of GIS relied on the link between a desktop GIS client to data stored within the database. The pivot to web-based GIS can be tracked to the release of ArcGIS Server in 2008 (ArcGIS version 9.0.1). This release marked the beginning of the pivot to promoting and providing GIS data on the web, as opposed to solely desktop clients.

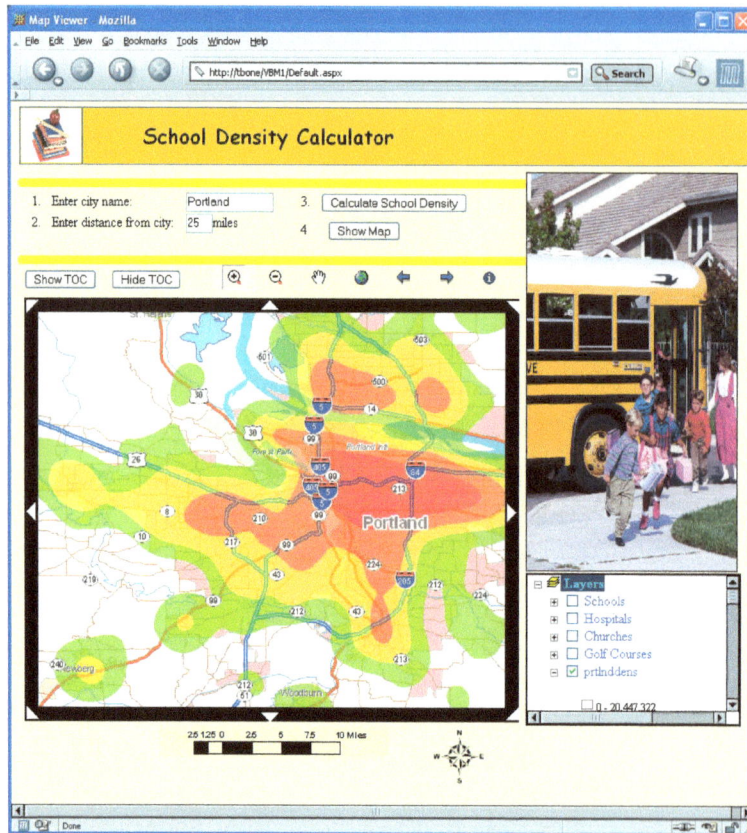

Figure 2.1. An early web page example of ArcGIS Server 9.0.1.

Up until the ArcGIS 10.5 release, ArcGIS Server was the only ArcGIS solution that would host user GIS data as web services. As a stand-alone solution, ArcGIS Server could
- Authenticate users (through built-in and organizational level users) and manage their access requirements,
- Host a variety of services, including map, image, and geocode services,
- Create connections to a variety of databases to provide an online editing experience,
- Offload certain workflows from the client ArcGIS Desktop software to the more powerful ArcGIS Server resources, and

2

- Scale by enabling users to add additional machines to an ArcGIS Server site, improving capacity and response times.

As an initial solution to web-based GIS, ArcGIS Server provided a simple framework for how to complete basic functions within the GIS world. However, it left some gaps in an industry that was quickly growing to encompass other vertices outside of GIS. For example, the lack of a built-in map viewer in ArcGIS Server forced any users who wanted to view or manipulate the data to have extensive knowledge of ArcGIS Desktop to make and maintain these connections. Additionally, there wasn't a clean solution to share data outside of the ArcGIS Server site without implementing complex geodatabase replication processes with other organizations. A change was needed to satisfy the growing demand in these areas.

Today's ArcGIS Enterprise was introduced at the release of ArcGIS 10.5. This release brought together what had previously been a separate product and extension—ArcGIS Server and Portal for ArcGIS—to a single unified product. ArcGIS Enterprise is made up of components including ArcGIS Server, Portal for ArcGIS, ArcGIS Data Store, and more. When combined with a functioning system, these components form a minimal ArcGIS Enterprise system, which is referred to as the ArcGIS Enterprise base deployment. Customers can expand on this system by adding additional capacity and functionality.

The ArcGIS Enterprise base deployment

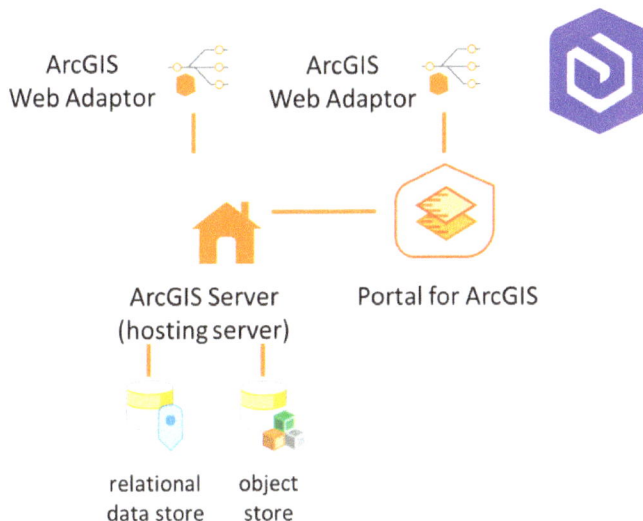

Figure 2.2. Logical diagram of an ArcGIS Enterprise base deployment.

This simple diagram shows the base deployment of ArcGIS Enterprise with four compo-nents—ArcGIS Server, ArcGIS Enterprise portal (referred to Portal for ArcGIS with respect to the installable component), ArcGIS Data Store, and ArcGIS Web Adaptor—representing the minimum configuration required for ArcGIS Enterprise to function. Let's take a close look at each of these components:

ArcGIS Server

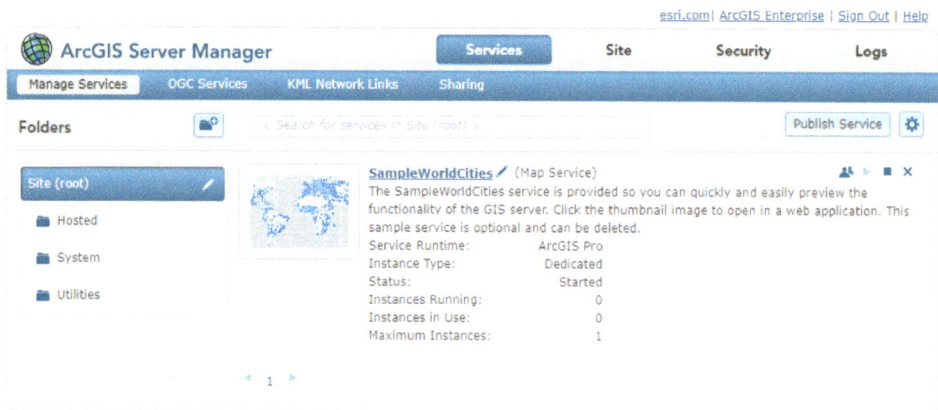

Figure 2.3. The ArcGIS Server Manager page.

ArcGIS Server is commonly referred to as the *engine* of ArcGIS Enterprise. As a stand-alone solution, ArcGIS Server enables access of services from a variety of data sources, such as enterprise geodatabases, file stores, and cloud stores. When federated as a hosting server, ArcGIS Server enables critical functions within the enterprise organization. ArcGIS Server can be used in stand-alone applications.

When accessing ArcGIS Server for the first time, you are presented with ArcGIS Server Manager. ArcGIS Server Manager displays key information about the services you have published, security details, site details, and logging information. We will explore the server manager in depth during the tutorial at the end of this chapter.

Although ArcGIS Server can be used as a stand-alone solution, ArcGIS Enterprise relies on ArcGIS Server to function as its service and data store manager. You need to become familiar with some terms to understand how this relationship works:

- **Federation**: Federation is the action of connecting an ArcGIS Server site to an ArcGIS Enterprise portal deployment. Federating an ArcGIS Server site to the ArcGIS Enterprise portal will have the following effects:
  - The ArcGIS Enterprise portal will now control security and access to the ArcGIS Server site.
  - If requirements are met, ArcGIS Server can be configured to a specific role to expand the functions of the enterprise organization.
- **ArcGIS Server role**: Depending on whether various prerequisites are met, ArcGIS Server can be configured to operate in different roles. The role of hosting server is required to complete a base deployment of ArcGIS Enterprise We will examine each of these roles in greater depth in chapter 5.
- **Services**: ArcGIS Server supports an array of web services based on the type of data being shared. If you consider a vector-based dataset, ArcGIS Server serves this data as a map service by default, with the option to enable feature access if editing is required. Any combination of service types can be enabled on a single service, depending on the user's need.
- **Data stores**: Data stores represent the sources of data that ArcGIS Server uses to serve different types of services. In this context, data stores may include databases, geodatabases, folders, and cloud stores. Data stores can also include the ArcGIS Data Store, which is a requirement for base deployment.

For ArcGIS Enterprise to function properly, you must register ArcGIS Relational Data Store to ArcGIS Server and have it elevated to the Hosting Server Role. This step is a minimum requirement for the seamless operation of ArcGIS Enterprise.

Note: The hosting server provides spatial tooling needed to conduct analysis within ArcGIS Enterprise. These tools exist as geoprocessing services and are used in many ways. For example, the same spatial analysis geoprocessing service is used to run the spatial analysis tools in Map Viewer and the spatial operations available within ArcGIS Pro (when configured with an ArcGIS Server connection).

The ArcGIS Enterprise portal

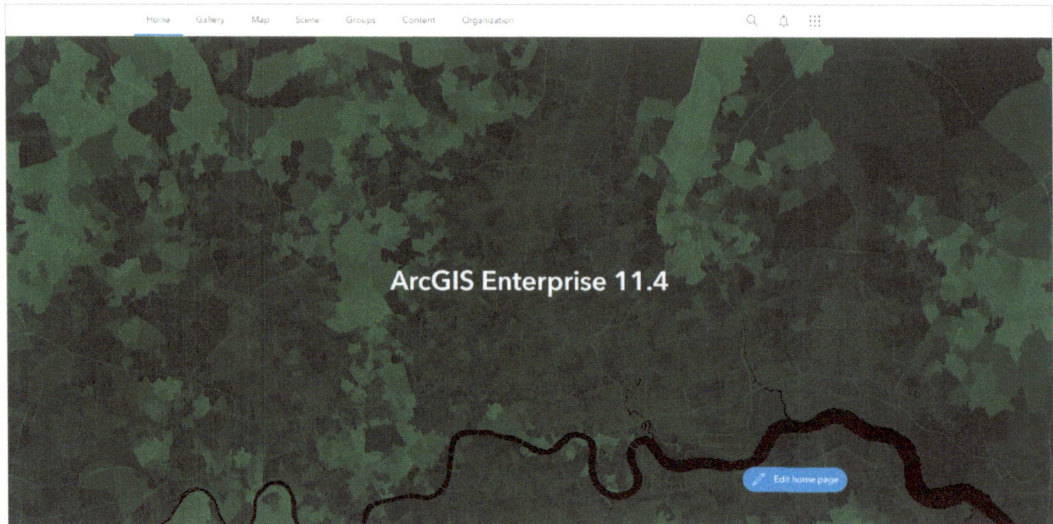

Figure 2.4. The ArcGIS Enterprise portal home page.

The ArcGIS Enterprise portal is a component in ArcGIS Enterprise that plays a foundational role in organizing and sharing information within an organization. In most cases, your users will interact with ArcGIS Enterprise through the ArcGIS Enterprise portal. Enterprise administrators manage their content with the users of an organization, while also including the ability to directly work with, display, and analyze the data contained within. It's helpful to think of the portal as a website that can access, manage, and display content among many users.

The ArcGIS Enterprise portal must have a federated ArcGIS Server that has been promoted to the hosting server role to function fully. You may notice that we refer to the ArcGIS Enterprise portal as Portal for ArcGIS from time to time. Portal for ArcGIS is the installed component on you server hardware. When it is being used from the browser or through any client, it is referred to as the ArcGIS Enterprise portal. When federated into the base deployment, the portal will support publishing services that power multiple built-in applications, such as ArcGIS Experience Builder and ArcGIS Instant Apps. The portal will also use geoprocessing services on the hosting server to power operations on the site. Think of the last time you published a feature layer or printed a web map. Both of those actions rely on a different geoprocessing utility to function. There are more than 25 distinct services that power different types of functions and analysis that occur on the portal.

> *Note: When federated with the ArcGIS Enterprise portal, ArcGIS Server no longer controls its own authentication and security. Administrators may use the portal to publish new data store connections, manage services, and publish geoprocessing services. For more information on the ArcGIS Enterprise portal, see chapter 6.*

The ArcGIS Data Store

The ArcGIS Data Store is the part of ArcGIS Enterprise that manages data storage. It is responsible for storing different types of data associated with hosted feature layers, tile caches, hosted scene layers, and more. The ArcGIS Data Store is configured with the hosting server in Windows and Linux deployments. For a base deployment to be complete, the ArcGIS Data Store must be configured with a relational data store and either a tile cache data store (ArcGIS Enterprise 11.3 and older) or an object store (ArcGIS Enterprise 11.4 and later).

The ArcGIS Data Store supports five configurations:

- **Relational data store**: The relational data store supports the creation and storage of hosted feature layers. This is a core requirement of a base deployment of ArcGIS Enterprise.
- **Spatiotemporal big data store**: The spatiotemporal big data store (STBDS) stores large datasets that include both space and time components and is designed specifically for high-volume and high-velocity data operations.
- **Tile cache data store**: The tile cache data store represents a set of databases that support hosted scene layers and stores tile caches before ArcGIS Enterprise 11.4.
- **Graph store**: The graph store is used with the ArcGIS Knowledge Server extension. It stores and manages the nodes and relationships that make up a knowledge graph.
- **Object store**: This store is responsible for storing cached query responses for feature layers (for performance improvement purposes) as well as video for a federated ArcGIS Video Server. After ArcGIS Enterprise 11.4, the object store supports hosted scene layers and hosts tile caches.

ArcGIS Web Adaptor

The ArcGIS Web Adaptor forwards requests to either ArcGIS Server or Portal for ArcGIS. Web Adaptor relies on a web server to function—for example, Microsoft's IIS or Java EE servers, such as WebSphere and WebLogic. Two web adaptors are required to be configured in a base deployment of ArcGIS Enterprise—one for the ArcGIS Enterprise portal and one for ArcGIS Server. In later chapters, we will discuss federating additional ArcGIS Server sites and setting up more complex deployments.

Fictional user story

Becken Pond Conservation Society is responsible for understanding how invasive species affect the microenvironment that surrounds small ponds and lakes. As a nonprofit that employs limnologists, biologists, and other scientists, it relies on fund-raising from surrounding communities to advocate for the well-being of these microbiomes.

Linda Jackson, the executive director of the Becken Pond Conservation Society, realized that she needs a way to track collection work and analysis around the state's ponds and create maps of potential neighborhoods that would benefit from the work that the society advocates for.

Linda hires Jim Yazzie, a GIS expert from the county, to design a system that can create, store, and share information about the ongoing work around these sites and generate maps of neighborhoods that may offer help to the society. As they develop the system plan, Linda and Jim realize that they could also store donor information in the system to better understand where the funding is coming from.

To ensure donor privacy, Jim proposes using a base deployment of ArcGIS Enterprise. He works on deploying ArcGIS Enterprise using ArcGIS Enterprise Builder on their internal web server. Upon patching the deployment with all available patches, Jim begins to implement the workflows defined by Linda in the early stages of the planning process. He also implements a backup strategy to ensure there is a save point for the deployment that can be relied on in case of any issues.

Tutorial 2: Explore the ArcGIS Enterprise portal and server manager

Explore the ArcGIS Enterprise portal in a base deployment

Let's start with a basic tour of your ArcGIS Enterprise deployment and verify that your ArcGIS Enterprise portal has been deployed and configured to run a base deployment of ArcGIS Enterprise, including its key components.

1. On the ArcGIS Enterprise home page, sign in as an administrator user.

2. On the top navigation bar, click the **Organization** tab.

    You will see an overview of your organization's information, including member details, recent content, and available add-on licenses.

3. Click the **Settings** tab. Then click the **Servers** tab to open configure server settings.

You should see one federated server with the hosting server role. Depending on your deployment, you may also see other federated servers.
- What is the Service URL of the hosting server?
- What is the Admin URL of the hosting server?

4. Hover over the **Status** value for the hosting server.

A pop-up appears with details about whether the hosting server has met validation requirements. Does the status indicate that every validation check has passed?

5. Repeat the status check for any other federated servers.

Explore ArcGIS Server Manager

The details described on the **Servers** tab of the ArcGIS Enterprise portal's **Settings** are all important to assess the stability of a federation. However, this information is a brief over-view. As an ArcGIS Enterprise administrator, you need to act on any messages of concern. To find out more information about any errors, you will need to access ArcGIS Server Manager.

6. In the list of **Federated server sites**, copy the **Service URL** and paste it to an empty tab of your browser. To the end of the URL, add /**manager** and press **Enter**.

> *Tip: For example, if your Service URL is https://gis.example.com/server, you can access the server manager at https://gis.example.com/server/manager. For more information on organization URLs, refer to the "Anatomy of an organization" URL section in chapter 6.*

7. If prompted to log in, since this ArcGIS Server site is federated, you can use the same administrator credentials used in the ArcGIS Enterprise portal for access.

The ArcGIS Server Manager landing page shows a list of all services running on this ArcGIS Server site. If this is the hosting server, you will see a folder for **Hosted**, along with the typical default ArcGIS Server folders for **System** and **Utilities**. Each of these contains an ArcGIS managed service, meaning that these services are supported by underlying data and processes that do not have direct access outside of the REST API.

8. At the top of the page, click the **Site** tab.

The default is opened to the **Directories** tab.

9. Click the **Machines** tab.
    - How many machines are included in this ArcGIS Server site?

10. Click the **Data Stores** tab.
    - Which data stores are registered with this ArcGIS Server site?

11. On the blue navigation bar, click the **Web Adaptor** tab.
    - Is Administrative Access enabled through the Server site web adaptor?

12. At the top of the page, click the **Security** tab.
    - Which ArcGIS Enterprise component controls the User Store for the ArcGIS Server site?
    - What is the username of the Primary Site Administrator (PSA)?

    Note: It is imperative that these security configurations are never modified after federation. Doing so may irreparably damage your ArcGIS Enterprise organization.

13. Return to the **Services** tab. Open the **Utilities** folder.

    These are some of the core tools that help run many applications within the ArcGIS Enterprise portal.
    - What does the Geocoding Tools service do?

14. On a separate tab, return to the ArcGIS Enterprise home page.

15. On the top navigation bar, click the **Content** tab. Then click the **My organization** tab.

16. In the search bar, search for **Tools**.

    You will see the same list of tools in the **Utilities** folder of ArcGIS Server Manager.

    As an ArcGIS Enterprise administrator, it is important to understand that most services that exist in the ArcGIS Enterprise portal have a corresponding service in ArcGIS Server. This relationship forms the backbone of distributed GIS, enabling your users to complete tasks quickly and efficiently.

Summary

In this chapter, we defined the ArcGIS Enterprise base deployment and explored the components that make up the greater whole.

CHAPTER 3
Deploying ArcGIS Enterprise

Objectives
- Describe the licensing requirements for ArcGIS Enterprise.
- Explain the benefits and limitations of various deployment options.
- Describe the process for installing a base deployment.

Introduction
The previous chapter focused on the conceptual understanding of an ArcGIS Enterprise base deployment. In this chapter, you'll learn the practical process for installing and configuring ArcGIS Enterprise on Windows and on Linux.

Licensing
Before installing any software, you should determine which ArcGIS Enterprise capabilities you need. The analysis of your business needs will inform your decision about which specific licenses you require to deploy those capabilities. The licensing model for ArcGIS Enterprise consists of two types of licenses.

A portal license file contains information about the capabilities you can assign to users. The number and kind of licensed user types, user type extensions, and any add-on licenses are included in this file. Chapter 7 has more information about the specific capabilities of different user types.

An ArcGIS Server authorization file contains information about the capabilities of an ArcGIS Server site. Different ArcGIS Enterprise server roles have different licenses, and some roles have separate licensing for advanced capabilities. You will need a separate authorization file for each ArcGIS Server site you deploy. In this chapter, we will focus on the Hosting Server role required for a base deployment. Chapter 5 covers additional server roles.

The ArcGIS Server authorization file also determines the maximum number of central processing unit (CPU) cores you are authorized to use for the ArcGIS Server site. For example, installing ArcGIS Server on a four-core physical machine requires a four-core license. For virtualized or cloud deployments, the correspondence is not as straightforward, because virtual cores (vCPU) generally represent hyper-threads rather than a full physical core. You must calculate the actual physical core equivalent to determine the license you need. For example, you need a four-core license if ArcGIS Server is installed on a virtual machine with 8 vCPU that is running on a host with four physical CPU cores.

Deployment process

For each component of ArcGIS Enterprise, the deployment process consists of two separate steps: installation and configuration.

1. **During installation**: You decide the directory of the machine file system where the software will be installed and which operating system (OS) account the software will run on.
2. **During configuration**: You set up the capabilities of the software. You will also decide the directories where the information needed by each component will be stored.

This two-step process matters because some installed software can be configured in multiple ways to provide different components. For example, each ArcGIS Server Site and Portal for ArcGIS requires its own installation and configuration of the web adaptor. On the other hand, a single installation of the ArcGIS Data Store software can be configured to provide multiple data store components to satisfy the minimum requirements for a base deployment.

Table 3.1. Installation requirements

Installed software	Configured base deployment component	Capabilities
Portal for ArcGIS	ArcGIS Enterprise portal	User and content management system
ArcGIS Web Adaptor	ArcGIS Enterprise portal Web Adaptor	Reverse proxy/load balancer that routes requests to the ArcGIS Enterprise portal
ArcGIS Server	Hosting server site	Exposes GIS resources as web services
ArcGIS Web Adaptor	Hosting server web adaptor	Reverse proxy/load balancer that routes requests to the hosting server site
ArcGIS Data Store	Relational data store	Data storage for hosted feature layers
	Tile cache data store	Data storage for hosted scene layers at version 11.3 and earlier
	Object store	Data storage for hosted scene layers at version 11.4 and later

Operating system account considerations

The most impactful choice you make during the installation step is determining the operating system account under which the software will run. Your choice matters because this account must have access to all the directories the software needs to read from or write to. You do not have to use the same account for each component, but it will generally be easier to manage ArcGIS Enterprise if you do.

Windows

When you install ArcGIS Enterprise software on Windows, you can use three types of accounts:

- A local account is constrained to a single machine and cannot access resources outside that machine. By default, the installation process will create a local account named "arcgis" for the software to run under.
- A domain account is centrally managed by your domain. You can use the same domain account across multiple machines.
- A group-managed service account (gMSA) is a special type of domain account that cannot be used for interactive logins and is constrained to a predefined group of machines.

A local account is the default because the software installer cannot assume you have a domain controller that manages domain accounts or gMSA. Local accounts, however, can be more challenging to use because they lack access to resources outside a single machine. It will be much easier to ensure ArcGIS Enterprise has access to the resources it needs if you use a domain account or gMSA. In addition, using a gMSA is a recommended security best practice.

Linux

When you install ArcGIS Enterprise software on Linux, you do not specify an account for the software to run under. Whichever user runs the installation process will be the user the software runs under. For that reason, you must have created the user account beforehand and ensure it has file access permission to all the directories where the software will read from and write to. In a multimachine environment, the user ID of this account should be the same across all machines. As a security measure, this account cannot be the root account.

Networking considerations

The various components of ArcGIS Enterprise use a variety of ports to communicate, and you will need to ensure that your network is configured to allow communication on those ports. There are three main causes of network communication difficulties to account for:

- **Firewall configuration**: If the ports used by a component are not open on the firewall, traffic cannot be routed to the component.
- **Port conflict**: If another application on the machine is already listening on the same port, it cannot be used by the ArcGIS Enterprise component.
- **Antivirus configuration**: Some antivirus applications will block traffic on some ports, even if the port is open on the firewall.

You should refer to the documentation for a complete list of ports used, but the most important ones are those used for communication with other components.

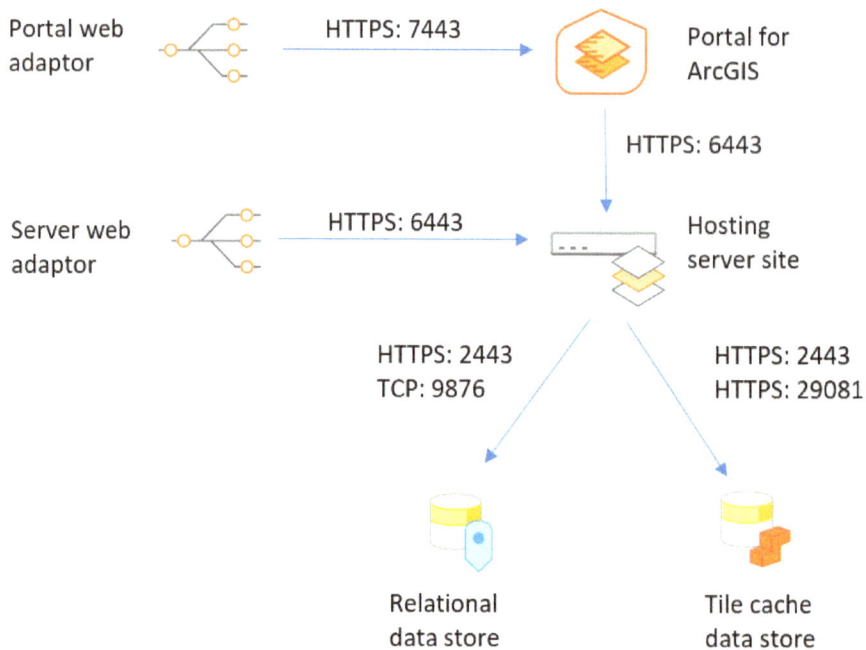

```
Portal web          HTTPS: 7443          Portal for
adaptor                                  ArcGIS

                                         HTTPS: 6443

Server web          HTTPS: 6443          Hosting
adaptor                                  server site

         HTTPS: 2443                      HTTPS: 2443
         TCP: 9876                        HTTPS: 29081

         Relational                       Tile cache
         data store                       data store
```

Figure 3.1. Ports used for communication between various ArcGIS Enterprise components.

Port numbers used by ArcGIS Enterprise are not user configurable. For example, if another application is already using port 6443, you will either need to change the port of that application or ensure that it runs on a separate machine from ArcGIS Server. You cannot fix the conflict by changing the port used by ArcGIS Server.

Deployment considerations

This is a good time to pause and consider how you want to deploy ArcGIS Enterprise.

Deployment order

The order in which you install and configure the components of ArcGIS Enterprise can matter. This is the recommended order:

1. Install Portal for ArcGIS.
2. Configure the ArcGIS Enterprise portal.
3. Install ArcGIS Web Adaptor.
4. Configure the web adaptor with the ArcGIS Enterprise portal.
5. Install ArcGIS Server.
6. Configure and license ArcGIS Server as an ArcGIS GIS Server site.
7. Install ArcGIS Web Adaptor again.
8. Configure the web adaptor with the GIS Server site.
9. Federate the GIS Server site with ArcGIS Enterprise.
10. Install ArcGIS Data Store.
11. Configure ArcGIS Data Store as a relational data store registered with the GIS Server site.
12. Configure the GIS Server site as the hosting server for ArcGIS Enterprise.
13. Install and configure ArcGIS Data Store for any additional data stores and register them with the hosting server.
14. Install and configure any additional ArcGIS Server sites.

The most common reason to deviate from this order is because you already have an existing stand-alone ArcGIS Server site that you want to integrate into ArcGIS Enterprise as the hosting server. That is a supported pattern, but for all other cases, you should follow the recommended order.

The ArcGIS Enterprise portal content directory

There are a few considerations related to the portal content directory that you specify when installing Portal for ArcGIS. The most important is that the OS account you specify has read-and-write permission on this folder. The account will have the right permission if you use the default location for the content directory, but other locations will require you to ensure the correct file system permission for the account.

For performance reasons, it is best to choose a location for the portal content directory on the local file system of the machine where Portal for ArcGIS is installed. If the system needs to access the information over the network, that will introduce latency that slows down the process. If you are configuring Portal for ArcGIS in a high-availability (HA) configuration, you may need to use a network location for the portal content directory, but otherwise it is generally better to use the local file system.

The portal content directory will also see a substantial number of reading and writing operations. Consequently, it is a good idea to choose a storage medium that is optimized for disk input/output (I/O) and can expand in size as more content is added.

Hosting server site

Like the portal content directory, there are some important considerations for the server directories and configuration store locations that you specify when creating the server site. Make sure that the ArcGIS Server account has access to these locations and use the local file system unless you are creating a multimachine ArcGIS Server site. These directories will not perform as many reading and writing operations as the content store, so it is less important to choose storage optimized for disk I/O, but faster disks will still see improved performance.

The minimum system requirements may not be sufficient for your needs if you will be publishing many services to the server site. Map and feature services are most likely to be memorybound, whereas geoprocessing services may be either CPU or memory bound. Be aware of the available RAM and CPU on the machines where ArcGIS Server is installed. Virtualized environments can help scale vertically by easily adding more memory or CPU, but that is more challenging when you must install hardware on a physical machine. You can also scale an ArcGIS Server site horizontally by joining additional machines to the site to add more computing resources.

ArcGIS Data Store

One consideration for data stores configured with ArcGIS Data Store is the amount of available disk space. If the available disk space comes close to the maximum, the data store will go into read-only mode and no changes can be made to any service that uses that data store. Because you may not be able to fully predict the total amount of disk space you will need for all your services, it is a good idea to use a storage solution that can be expanded, as necessary.

A second consideration for ArcGIS Data Store is that the initial backup configuration is not appropriate for a production environment. For example, ArcGIS Data Store configures the backup location of the relational store in the same directory as the data itself. This happens because that location is the only one that the installation software can be sure has the right permission for the account the ArcGIS Data Store software is running under. But software defaults are different from best practices. After installing ArcGIS Data Store, you should configure the backup location to a location on a separate machine so that you won't lose the backup if the machine goes down. You should also schedule data store backups to meet your organization's recovery time and recovery point objectives.

Specific deployment options

This section reviews options for deploying ArcGIS Enterprise using ArcGIS Enterprise Builder, individual components, Infrastructure as Code (IaC), and the cloud.

3

ArcGIS Enterprise Builder

The easiest way to deploy ArcGIS Enterprise is to use ArcGIS Enterprise Builder, an installation and configuration wizard that sets up a base ArcGIS Enterprise deployment on a single machine. In this option, you specify a few configuration options, and then the software installs and configures a base deployment. This option takes about an hour to complete and doesn't require any monitoring or intervention once it has begun.

Despite its ease of use, the enterprise builder has key limitations that mean it is not a good choice for many production environments. Every component will be installed on a single machine, which makes it difficult to scale the deployment to handle more services than is possible on a single machine. You also have substantially less control over some configuration options, such as different OS accounts for different components, the location of the installation directories, or domain name system (DNS) aliases. The enterprise builder also does not automatically configure web adaptors in Linux environments.

The enterprise builder is a good choice for solutions that aren't affected by its limitations, such as quickly setting up a test environment. You can also deploy with the enterprise builder initially and then change the configuration later to provide more capabilities. Changing the configuration, however, means that you will not be able to use the enterprise builder to upgrade when a new version of ArcGIS Enterprise is released.

Fictional user story

Harjeet Singh, the SuperBiz director of data, has gotten some feedback from users that internal frontline support analysts are frequently not able to help troubleshoot challenges people face using ArcGIS Enterprise. Investigating the issue further, Harjeet learns that most analysts have never received formal training in administering ArcGIS Enterprise, and they have no access to test environments for replicating users' issues.

Because SuperBiz has enterprise-level agreements with both Esri and Microsoft, Harjeet knows he can deploy any number of ArcGIS Enterprise organizations on Windows. He coordinates a project to set up test deployments for analysts using ArcGIS Enterprise Builder, which simplifies the work of creating many identical deployments. Analysts use these test systems to get direct practice with ArcGIS Enterprise and test issues users are experiencing in the production environment. Going forward, when SuperBiz upgrades its production ArcGIS Enterprise environment, the enterprise builder simplifies the process for upgrading the analysts' test environments to match.

Individual components

A second option is to install each component of ArcGIS Enterprise separately. For this option, deployment order matters the most because you are responsible for ensuring the proper order. This is a good choice for organizations that have more complex needs that can't be met by the ArcGIS Enterprise Builder but don't have the DevOps culture or cloud services expertise to use a fully automated option. Installing individual components is also an excellent way to learn how the deployment process works. We recommend that every ArcGIS Enterprise administrator attempt this at least once in a test environment as a professional development exercise.

Fictional user story

The Becken Pond Conservation Society has long used its gis.bpcs.org machine to run ArcGIS Server and make its data available as web services. When it made the decision to deploy ArcGIS Enterprise, it wanted to reuse its existing infrastructure. Jim Yazzie, the BPCS GIS expert, knows that the society's ArcGIS Enterprise deployment will be small enough that it could fit on a single machine, but the requirement to reuse the existing ArcGIS Server site means that the ArcGIS Enterprise Builder was not a viable choice.

A conversation among some of the people who would use the new ArcGIS Enterprise deployment also indicated that they weren't clear about what "GIS" means. That wasn't an issue before because users weren't directly engaging with service URLs. But they would be navigating the ArcGIS Enterprise portal site. Jim decided to deploy the portal web adaptor on its own virtual machine named maps.bpcs.org to make its purpose clearer to users.

By installing the components individually, Jim was able to accomplish all the organization's goals. Reusing the existing ArcGIS Server site prevented the need to migrate services. Putting most components on a single machine reduced hardware costs. Having a better URL for users to access the ArcGIS Enterprise portal improved user awareness of the uses of ArcGIS Enterprise.

Scripted deployment

Organizations with strong DevOps experience and culture may want to automate the deployment of ArcGIS Enterprise using IaC. In this option, the authoritative source of truth for the correct configuration of ArcGIS Enterprise is held in one or more configuration files. A configuration management tool takes these files as inputs and reliably reproduces the ArcGIS Enterprise deployment.

Scripted deployment is convenient for situations in which you need to repeatedly deploy a complex configuration of ArcGIS Enterprise. The configuration scripts are also useful because they serve as documentation for the correct configuration that is separate from the deployment itself. For example, if the machine where you installed Portal for ArcGIS is down and you need to get the fully qualified domain name of every federated server, you can consult the configuration file you used to deploy the system in the first place.

3

ArcGIS Enterprise supports two script-based deployment frameworks: Chef and PowerShell DSC. Chef uses files called cookbooks written in the Ruby programming language that work across either Windows or Linux environments. PowerShell DSC is a Microsoft tool that uses the PowerShell language to manage configuration in Windows environments. Esri publishes both prewritten Chef cookbooks and PowerShell modules that you can use for a variety of deployment scenarios.

The major limitation of scripted deployment is that it requires experience with the configuration management tools. Many organizations are already using Chef or PowerShell DSC to manage other parts of their infrastructure, so it is easy for them to extend that work to ArcGIS Enterprise. For organizations without that experience, however, it will be challenging to use these tools effectively.

Cloud deployment

Any of the previous deployment options can be used with either cloud-based infrastructure or on-premises. But when deploying on Amazon Web Services (AWS) or Microsoft Azure public cloud platforms, specialized tools make it easier to deploy ArcGIS Enterprise. These tools come in two varieties: ArcGIS Enterprise Cloud Builder and configuration templates.

ArcGIS Enterprise Cloud Builder for AWS and ArcGIS Enterprise Cloud Builder for Microsoft Azure are apps that you install on your local machine. These apps guide you through the process of deploying ArcGIS Enterprise in their respective cloud environments. This guided process simplifies the experience and can be especially helpful for people who are new to deploying ArcGIS Enterprise.

The configuration templates use existing Infrastructure-as-a-Service (IaaS) offerings from both AWS and Azure. These configuration templates are JSON files that contain the information needed to deploy ArcGIS Enterprise. Esri publishes separate CloudFormation templates for AWS and Azure Resource Manager (ARM) templates for Microsoft Azure. Like Chef cookbooks and PowerShell DSC configuration files, these cloud templates work well for organizations that are already used to them but may be more challenging to use for other organizations.

Fictional user story

At around the same time that Medio County started exploring the deployment of ArcGIS Enterprise, the county executive announced a new partnership between the county and Microsoft, mandating that new IT infrastructure be deployed in Azure. Elise Medina, the county GIS manager, had always deployed infrastructure on premises. She wasn't sure how to proceed with deploying in the cloud, so she turned to Jake Nilsen, the county IT director, for help.

Elise and Jake immediately rejected the option to deploy individual components, because they wanted something that could simplify the process. Enterprise Builder was attractive for that reason, but Elise knew the county would likely outgrow a single-machine deployment quickly. Having to rearchitect ArcGIS Enterprise so soon after the initial deployment didn't appeal to either Elise or Jake, so they rejected the Enterprise Builder option.

Jake's team used a configuration management tool called Ansible to automate some processes already, so he suggested they explore the scripting and templating options available for ArcGIS Enterprise. They were able to get a working ArcGIS Server site deployed using an ARM template. Because neither Jake nor Elise had used Chef, PowerShell DSC, or ARM templates before, it was difficult to understand how they worked. They weren't willing to rely on a tool they didn't fully understand, and training on these tools would take longer than they wanted. Jake thought he could probably create something using Ansible, but without premade configuration playbooks, it would be a significant undertaking. For those reasons, they ended up rejecting the option to use scripts or templates as well.

Finally, they tried the ArcGIS Enterprise Cloud Builder for Microsoft Azure. This simplified the process in a way that didn't require deep preexisting expertise in the tool. Jake and Elise were confident that they could extend and upgrade ArcGIS Enterprise when needed, so they deployed both their test and production environments using this option.

Tutorial 3: Explore your ArcGIS Enterprise deployment

Before you start to perform the tutorial steps, carefully consider these questions, which will help you prepare for deploying ArcGIS Enterprise and make deployment decisions.

If you do not already have ArcGIS Enterprise deployed, use the questions to help you think through the choices you will make when you deploy ArcGIS Enterprise. Many of these questions are predicated on a thorough understanding of your requirements and will likely require collaboration across different parts of your organization to answer fully. That may not be easy, but you should do it before you deploy ArcGIS Enterprise. Otherwise, you run a strong risk of deploying a system that doesn't meet your needs.

If you already have ArcGIS Enterprise deployed, the tutorial will guide you through exploring how it has been deployed. First, you will want to consider these sets of questions to achieve optimal results.

3

System patterns, licenses, and server capabilities

- What system patterns does your ArcGIS system need to support? (Refer to the ArcGIS Architecture Center for more information on system patterns.)
- What user type licenses will you need to provide users? (See chapter 7 for more information on user types.)
- What ArcGIS Server capabilities do you need? (Refer to the ArcGIS Enterprise functionality matrix for more information about ArcGIS Server capabilities.)

OS account

- Are you going to deploy on Windows or Linux?
- If Windows, do you have a domain or gMSA account you can use to run the software under?
- Which directories do you plan to use for the software installation files?
- Which directories do you plan to use for the data, content, and configuration files used by different ArcGIS Enterprise components?
- Does the account you plan to use have read-and-write access to these directories?

Networking

- Who in your organization will be responsible for ensuring that the correct ports are available for ArcGIS Enterprise?
- How will you document the ports used by ArcGIS Enterprise so that subsequent security audits don't close them?

Choose a deployment option

- Are you going to install all ArcGIS Enterprise components on a single machine?
- If so, is it appropriate to use Enterprise Builder?
- Do you plan to deploy on either AWS or Azure cloud environments?
- If so, is it appropriate to use the ArcGIS Cloud Builder templates?
- Do you need an IaC pattern for deploying ArcGIS Enterprise?
- If so, does your organization already have expertise with either PowerShell DSC or Chef/CINC?

- If your organization lacks the needed expertise, how will you provide for learning and practice opportunities?
- If your organization has a business continuity and disaster recovery plan, does this deployment option meet the requirements of that plan?

Backups

- What is your necessary recovery time objective (RTO)?
- What is your necessary recovery point objective (RPO)?
- What kind of backup schedule will you set to meet your RTO and RPO?

Getting started

If you already have ArcGIS Enterprise deployed, you can follow the steps in this tutorial to explore how it was installed and configured. Performing all the tasks in this tutorial requires the ability to access information on the machines where ArcGIS Enterprise components are installed.

Licensing

1. In a browser, sign in to your ArcGIS Enterprise portal organization as an administrator.

   Tip: For more information on the ArcGIS Enterprise portal interface, refer to chapter 6, "Getting Familiar with the ArcGIS Enterprise Portal."

2. On the navigation bar, click the **Organization** tab and then click **Licenses**.
   - How many licenses do you have for each user type?
   - When do these licenses expire?

3. Click **Settings** and then, on the left, click **Servers**.

4. For the server site with the **Hosting Server** role, click the three horizontal dots (More Options) next to the name. Then click **View licenses**.

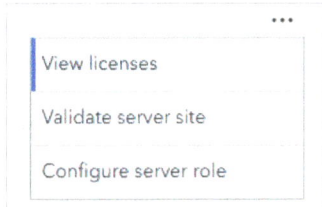

- What role is your hosting server site licensed for?
- When does the role license expire?
- What extensions, if any, is your hosting server site licensed for?
- When do those extensions expire?

Machine used by Portal for ArcGIS

You will examine the details of the machine used by your ArcGIS Enterprise portal in the Portal Administrator Directory.

1. Copy the current URL on the browser tab of your ArcGIS Enterprise portal organization page.

2. On a new browser tab, paste the copied URL. In your URL, replace everything after **portal** with /**portaladmin**. Press **Enter**.

   Tip: For example, if your portal organization page is gis.example.com/portal/home/organization, your Portal Administrator Directory is gis.example.com/portal/portaladmin.

3. Sign in to the Portal Administrator Directory as an administrator. You can use the same administrator credentials used in ArcGIS Enterprise for access.

- What version of Portal for ArcGIS is running?

4. From the **Site Root**, within **Resources**, click **Machines**. Then click the name of one of the machines listed.
   - What is this machine's name?
   - Is it running Windows or Linux?

5. At the top of the page, click **Home** to return to the Portal Administrator Directory Site Root.

6. From **Resources**, click **System** and then click **Directories**. From the list of directories, click **content** to list information about the portal content directory.
   - What is the file path for the content directory?

Machine used by ArcGIS Server

1. At the top of the page, click **Home** to return to the Portal Administrator Directory Site Root.

2. From **Resources**, click **Federation** and then click **Servers**. Click the link for the hosting server.

3. Under the hosting server information, copy the URL and paste it into a new browser tab.

4. Change the URL by adding **/admin** to the end. Press **Enter**.

   This opens the ArcGIS Server Administrator Directory.

   > *Tip: For example, if your hosting server URL was https://gis.example.com/server, your Server Administrator Directory is https://gis.example.com/server/admin.*
   >   *If you are not able to access the Server Administrator Directory at that URL, try using the Admin URL listed for the hosting server instead—for example, https://gis.example.com:6443/arcgis/admin.*

5. Sign in to the Server Administrator Directory using either Primary Site Administrator credentials or follow the directions on the page to sign in with an OAuth portal token.

ArcGIS Server Administrator Directory

Home

You should use ArcGIS Server Manager for managing services and GIS servers.
The Administrator Directory is intended for advanced, programmatic access to the server, likely through the use of scripts.

Site Root - /

Current Version: 11.4.0

Resources: machines services security system data uploads logs kml info mode usagereports publicKey

Supported Operations: generateToken exportSite importSite deleteSite

Supported Interfaces: REST

---

6. From the Server Administrator Directory **Site Root**, within **Resources**, click **machines** and then click the name of one of the registered machines listed.
   - What is the name of this machine?
   - Is it running Windows or Linux?

7. At the top of the page, click **Home** to return to the Server Administrator Directory Site Root.

8. Click **system**, then click **configstore** to list information about the location where the essential site properties are stored.
   - What is the location of the configuration store?

9. At the top of the page, click **system** and then click **directories**. Select **arcgisinput** to list information about the directory where source files for published services are stored.
   - What is the file path of the arcgisinput directory?

Machine used by the relational data store

1. At the top of the page, click **Home** to return to the Server Administrator Directory Site Root.

2. Click **data** and then click **items**. From the list of **Root Items**, select **/enterpriseDatabases**.

3. Under **Child Items**, click the link for a child item with **AGSDataStore** in the name.

   This is a relational data store.

4. Click **machines** and then click the name of the machine.
   - What is the name of this machine?
   - Is it running Windows or Linux?
   - Is the role of this machine configured as the primary relational data store?

OS account

1. Sign in to the machine where Portal for ArcGIS is running that you identified earlier.

   > *Tip: Depending on your organization's security settings and the machine's OS, there are a variety of ways to sign in to the machine. Work with your organization's IT staff if you are not sure how or whether you can access this machine.*

2. If the machine is running Windows, open the **Services** app from the **Start Menu**.

3. If the machine is running Linux, there isn't a universal method that will work for all supported configurations, but the most likely method to succeed is running the following command from the terminal:

   ```
   pgrep arcgis | xargs ps -u -p
   ```

4. For Windows, find the name listed in the **Log On As** column for the Portal for ArcGIS service. For Linux, find the name listed in the **USER** column for Portal for ArcGIS.
   - What is the OS account name used by Portal for ArcGIS?

5. Repeat the process for machines where the hosting server site and relational data store are running.
   - Does ArcGIS Server use the same OS account name as Portal for ArcGIS?
   - Does ArcGIS Data Store use the same OS account name as Portal for ArcGIS?

Backups

1. Sign in to the machine where the relational data store is installed and open a command line terminal with administrator privileges.

   > *Tip: Because there is no GUI for managing ArcGIS Data Store, you will use the command line utilities in the terminal.*

2. In the terminal, navigate to the ArcGIS Data Store installation directory. By default, this is the following directory:
   - For Windows: `C:\Program Files\ArcGIS\DataStore`.
   - For Linux: `$HOME/arcgis/datastore` where `$HOME` is the home directory of the OS account that ArcGIS Data Store is running under.

   > *Tip: For deployments on Ubuntu, it is common for ArcGIS Data Store to be installed at `/opt/arcgis/datastore` and the data to be located at `/gisdata`.*

3. From the installation directory, navigate to the **tools** directory.

4. Execute the utility to describe the data stores running on this machine using the following command:
   - Windows: `describedatastore`
   - Linux: `./describedatastore.sh`

5. Review the information for the relational data store.
   - What is the file path for the backup location?
   - What is the backup schedule?
   - How long are backups retained?
   - Is point-in-time recovery enabled?

6. Execute the utility to list the available backups using the following command:
   - Windows: `listbackups`
   - Linux: `./listbackups.sh`

7. Find the information for the most recent backup taken.
   - When was the backup taken?
   - What is the status of the backup?

Take the next step

If you do not already have ArcGIS Enterprise deployed, use the information in this chapter and in the documentation to create a base deployment.

Summary

In this chapter, you learned the process for deploying ArcGIS Enterprise. That process includes considerations you should account for prior to deploying, such as the licenses you require or system architecture you should design. It also includes choices about the mechanisms for deploying ArcGIS Enterprise, such as individually installing each component or using automated cloud deployment options.

CHAPTER 4

Introduction to ArcGIS Enterprise security best practices

Objectives

- Understand the ArcGIS Enterprise ownership-based security model.
- Explore software-level options to improve security.
- Exhibit Esri's commitment to security through the ArcGIS Trust Center.

Introduction

As enterprise-grade software, ArcGIS Enterprise is designed to ensure a secure user experience. ArcGIS Enterprise provides a built-in content management system that enables users to safely collaborate with each other in the context of the ArcGIS Enterprise portal. Although not a comprehensive guide to implementing security, this chapter is a primer to security concepts that apply to ArcGIS Enterprise.

The goal in any security strategy is to find a balance between the functionality needed to support the organization mission and security policies that limit unnecessary functions. Finding this balance will require negotiations between security stakeholders at your organization and GIS teams, so it is best to come prepared to this discussion with a clear set of workflows that need to be implemented.

In the scope of ArcGIS Enterprise, there are three primary areas where IT and ArcGIS Enterprise administrators should work together to establish common ground in their security practices:

- ArcGIS Enterprise ownership-based access security
- Software-level security
- Network- and peripheral-level security

This chapter does not include an exhaustive list of security considerations that must be implemented to secure a system. The next few pages instead will present basic information and describe a security structure whereby an ArcGIS Enterprise administrator can facilitate discussions with their IT staff to make security decisions. For more technical information, review the content in the ArcGIS Trust Center, a resource maintained by Esri's dedicated product security and privacy team. This chapter will conclude with a tutorial on how you can validate your system security with built-in tooling, using the ArcGIS Security and Privacy Advisor linked from the ArcGIS Trust Center.

ArcGIS Enterprise ownership-based security

This section covers these facets of security:

- Users, groups, roles
- Authentication options: built-in, organization-specific logins (web-tier, SAML, LDAP, OIDC, and so on)
- Audit logging

Users, groups, roles

ArcGIS Enterprise uses an ownership-based security model. All content within your organization is managed and secured through users and groups. Unless content is shared publicly, authentication is required to access and interact with most of the apps, features, services, and maps within ArcGIS Enterprise. Each user may be granted specific licenses, roles, and privileges to customize their ability to manage their own content within ArcGIS Enterprise. Even administrator-level privileges, such as creating groups and managing group content export, can be delegated to nonadministrators for delegation of responsibilities. We will explore content and user management in chapters 7 and 8, but let's talk about how ArcGIS Enterprise can use different authentication methods.

Authentication options

ArcGIS Enterprise supports creating and managing users using built-in authentication as well as an organization's identity stores, such as Active Directory, Lightweight Directory

Access Protocol (LDAP), Security Access Markup Language (SAML) compliant, or OpenID Connect (OIDC) identity providers. This capability allows administrators a certain level of control over how their users will authenticate against ArcGIS Enterprise. Note that no matter what level of authentication you choose, core functions, such as group creation and licensing, will be unaffected. If you are an application developer, ArcGIS Enterprise also supports noninteractive means of authentication using API keys and exchanging ClientIDs and Client Secret values. Let's examine the primary difference between built-in authentication and centralized authentication.

Built-in: This is the default identity store for your ArcGIS Enterprise portal. Upon creation, ArcGIS Enterprise prompts you to provide a built-in site administrator account. This account is used for the initial creation of the ArcGIS Enterprise portal. Although this is an acceptable way to start creating your ArcGIS Enterprise deployment, it is industry best practice to centralize account management through an organization-wide identity provider.

Centralized identity management (CIM): As a best practice, having a user base that is using CIM in an organization is critical toward establishing zero trust architecture (ZTA). The ArcGIS Enterprise Hardening Guide, which you can access at links.esri.com/GTKEnterprise-HardeningGuide, defines ZTA as the principle that "no actor, system, network, or service operating outside or within the security perimeter is trusted." CIM through an SAML-compliant internal developer platform (IDP) or other option aligns with ZTA foundational tenants.

Once ArcGIS Enterprise administrators choose and implement an authentication method, they must choose the roles and user types needed for each user to complete their work. Follow this primary security principle: Only grant users the exact privileges they need to complete their work. As simple as this sounds, defining specific rights and privileges to your default users as well as your power users is one of the most important ways to avoid accidents and spills in the future of your environment.

Next, consider the administrator user: A built-in administrator user is created at the beginning of the ArcGIS Enterprise portal installation. It is vital for the initial system configuration of ArcGIS Enterprise; however, it is also viewed as a potential security vulnerability because of its unassigned status as well as its elevated privileges. The ArcGIS Enterprise Hardening Guide recommends that this user is deactivated as soon as possible to avoid unnecessary exposure. Using custom roles, nonadministrator users may be delegated certain administrative rights and privileges. Custom roles allow organizations to better segregate administrative responsibilities across the ArcGIS Enterprise organization.

A good example of this is the Group Manager role. In the older releases of ArcGIS Enterprise, it was necessary to be an administrator or a group owner user to add, remove, and promote different users within a group. This task proved to be cumbersome to larger organizations that had many groups, so the Group Manager role was introduced. As a delegated role, nonadministrators and group owners had the ability to accept membership requests and manage the group. Delegating administrative responsibilities is a smart way to limit the number of administrator users in ArcGIS Enterprise for better security and accountability.

Audit logging and SIEM

Audit logging was introduced in ArcGIS Enterprise 11.4. Audit logging is the ability to track and report information related to

- access,
- item edits,
- item management,
- other key administrative tasks within ArcGIS Enterprise.

Audit logging is particularly useful to organizations that have legal requirements to track each change made to their data.

For example, let's say that you are an insurance provider. Since details surrounding users' claims can be considered privileged information, your organization may require a chain of custody to certain pieces of content within ArcGIS Enterprise. With audit logging, you can honor this requirement by tracking user sign-in information, item access and modification history, and group management information.

Audit logs captured in ArcGIS Enterprise are designed to be used in a Security Information and Event Management (SIEM) system. If a security event occurs, the SIEM allows organizations to access logs and events that occurred in the system at a specific time to correlate and analyze responses. Additionally, logs at various components of ArcGIS Enterprise may be directed into the SIEM to understand the software-level events that may have contributed to an event.

This chapter covers two major topics when you consider application-level security: user authentication and auditing. Regardless of your implementation plan, these factors are important to consider early on. For details that are more nuanced toward a specific architecture or deployment style, consult the ArcGIS Enterprise Hardening Guide, found at the ArcGIS Trust Center. This guide covers the information we've presented here with added depth and context.

Software-level considerations

So far, we've talked about some features and functions that exist within ArcGIS Enterprise as an application. Because the software can be deployed on multiple machines and used as essential infrastructure, administrators must ensure that ArcGIS Enterprise is running in the most secure state possible. At a minimum, ArcGIS Enterprise administrators should be aware of these aspects:

- ArcGIS Enterprise versioning implications
- Certificates
- Security validation tooling available in ArcGIS Enterprise

Versioning implications

Esri typically releases two versions of ArcGIS Enterprise each year—a long-term supported (LTS) release and a short-term supported (STS) release. Deciding on which release to install depends on how often you want to upgrade. We will explore more of this nuance in chapter 20, but briefly, if having the latest available features is important to your workflows, you will want to upgrade to the STS and LTS releases as soon as possible.

From a security posture, upgrading to the latest release ensures that the underlying components of ArcGIS Enterprise are their most recent versions. Due to different dependencies, Esri does not recommend upgrading these components independently. Doing so will lead to unexpected consequences and may cause the system to become unsupportable. Next, we'll review the necessary steps to secure communication between these components using certificates.

> Tip: ArcGIS Enterprise changed the way versioning is reported between the ArcGIS Enterprise 10.9.1 release and the ArcGIS Enterprise 11.0 release. Read this blog to learn more: "What's in a Number?" at links.esri.com/GTKEnterprise-VersionNumbering.

Certificates

What are certificates? In the context of enterprise networking, an enterprise network certificate, also known as an Enterprise SSL Certificate/Enterprise PKI Certificate, is a way to establish trust within a network. Depending on how certificates are created, they can be countersigned by a Certificate Authority (CA). A CA is responsible for creating a trusted list of root certificates that support the basic foundation of Hypertext Transfer Protocol Secure (HTTPS) encrypted communication.

You can import different kinds of certificates into ArcGIS Enterprise to establish trusted communication between

- components in your deployment,
- distributed collaborations between other ArcGIS Enterprise or ArcGIS Online organizations,
- ingress of data sources and feeds.

In the scope of a base ArcGIS Enterprise deployment, certificates become relevant in three places:

- ArcGIS Web Adaptor
- Internal ports used for communication between ArcGIS Server and the Portal for ArcGIS
- ArcGIS Data Store

Certificates are typically applied to Web Adaptor during installation; however, administrators usually overlook importing certificates at the internal end points of Portal for ArcGIS, ArcGIS Server, and ArcGIS Data Store. Although this will not influence your federation, it may have negative side effects when you work with advanced features, such as distributed collaboration, or use stream services from ArcGIS GeoEvent™ Server.

Portal for ArcGIS and ArcGIS Server components rely on communication over ports 6443 and 7443. Communication over these ports is vital for the internal functions of ArcGIS Server and the ArcGIS Enterprise portal, so you should never expect to see these ports in regular workflows. However, if certificates are not imported at these end points, you run the risk of unexpected failures in certain tasks occurring. Since the ArcGIS Data Store is tied to the hosting server, an administrator may choose to update the certificate using command line tools. This will secure communication between the hosting server and the ArcGIS Data Store over the specific ports. We will explore how to update all these certificates in the tutorial at the end of this chapter.

Security validation tooling in ArcGIS Enterprise

Administrators can use several tools to start the process of verifying the security of their ArcGIS Enterprise deployment. You will have the opportunity to work with these tools in the tutorial at the end of this chapter.

- **Built-in security tooling:** Portal for ArcGIS and ArcGIS Server include built-in scanning tools that administrators can use to quickly identify any unresolved security issues. These Python script tools are portalScan.py and serverScan.py, respectively. Each tool analyzes configuration properties and system details to display messages in three separate categories: Critical, Important, and Recommended. These issues all correlate with the ArcGIS Hardening Guide, and it's a good idea to have this available as a reference as you go through the results of these tools.
- **ArcGIS Security Adviser:** ArcGIS Security Adviser is an app-based tool that registers directly to ArcGIS Enterprise. It enables administrators to get a top-down view of their security status directly in the browser. No sensitive information is captured; however, users must register the ArcGIS Trust Adviser as a client-based web app.

Each of these tools provides a good starting point for an ArcGIS Enterprise administrator to understand any possible security gaps. However, the authors encourage all administrators to work with their IT and security teams on taking additional steps to ensure security. These steps can include creating governance policies for sensitive data and educating the user base on common security pitfalls.

Network considerations

Network-level security is dependent on the architecture adopted by your organization's IT department. There are innumerable variances in the way organizations configure their networking, ranging from the cloud provider they choose to the accessibility of the deployment outside the internal network. This chapter does not cover every possible nuance but rather serves as a guide to have conversations with your IT team to reach shared understandings and create a secure and accessible ArcGIS Enterprise deployment.

For a successful ArcGIS Enterprise deployment, it is vital that the IT and GIS departments align to create a conceptual plan. Too often, only one of the departments takes on the responsibility, which can lead to limited functionality, unexpected issues due to access problems, and potential data spills or leaks.

> Tip: To facilitate technical conversations and concepts, Esri has created the ArcGIS Architecture Center, links.esri.com/GTKEnterprise-ArchCenter. This resource defines the ArcGIS Well-Architected Framework, which demonstrates architecture best practices, provides an overview of ArcGIS as a platform as well as its deployment options, and brings additional training resources to the system that powers ArcGIS.

The elements that define your network security will vary based on the type of infrastructure on which you choose to deploy ArcGIS Enterprise. One common theme binds network security together, regardless of your infrastructure options: the level of availability of your content and ArcGIS Enterprise to the internet.

ArcGIS Enterprise is designed to run independently without any access to the internet. However, running a completely air-gapped deployment will also restrict your access to Esri's basemaps, ArcGIS Living Atlas of the World content, and other useful services. Earlier in the chapter, we said that the goal in any security strategy is to find a balance between the functionality that is needed to support the mission and a security policy that limits unnecessary functions. Although an air-gapped system may be the most protected from a security perspective, it comes at the cost of functionality. This is why understanding the type of data being hosted on your ArcGIS Enterprise deployment and the security requirements of the base data become so important.

To strike this middle ground, administrators often enable access to arcgis.com from their organizations to access ArcGIS Living Atlas content and basemaps. One of the most fundamental ways to secure this type of content access is using a web app firewall (WAF).

A properly configured WAF will monitor and filter all HTTP traffic to and from a web app. In the case of an ArcGIS Enterprise deployment, WAF rules will allow for the use of ArcGIS Living Atlas content and basemaps within the organization. The ArcGIS Trust Center's ArcGIS Enterprise Web Application Filter Rules documentation contains specific rules needed for the software to function properly.

> *Note: Esri takes security concerns seriously. The ArcGIS Trust Center allows any users of ArcGIS Enterprise to report a security or privacy concern. The Esri Product Security Incident Response Team is a dedicated group of security professionals who can verify incoming security concerns and respond in a timely manner. For more information, go to trust.arcgis.com.*

Fictional user story

SuperBiz has identified ArcGIS Enterprise as a potential candidate to provide traffic directions and fleet tracking of its road-based assets. Before deploying their development environment, the ArcGIS Enterprise administrator and IT department lead get together to understand the security requirements for the environment. First, the requirements of the projects need to be identified.

To properly create driving maps with full basemaps in ArcGIS Field Maps, a connection to ArcGIS Online would be needed.

SuperBiz was working on developing a custom locator to assist with route creation; however, for the proof of concept, the ArcGIS Routing Service may be used.

A vendor-provided data stream is used to monitor the truck's position and speed. This data stream is provided over a secure WebSocket.

Due to the sensitive nature of some of the data being used in SuperBiz's operation, the administrator carefully considers the type of data needed to fulfil the requirements identified earlier. The organization WAF is configured to allow necessary traffic through the firewall to ensure that a connection to ArcGIS Online's basemaps and routing service is achieved.

The vendor data stream uses WebSockets provided over a secure connection. The ArcGIS Enterprise administrator worked with the IT department to understand the origins of these requests and added their domain through the allow list.

After deploying their ArcGIS Enterprise deployment, the ArcGIS Enterprise administrator secured the deployment by enabling an organization-wide identity provider. This step imported premade user groups, including drivers that would need access to Field Maps, and different tracking capabilities offered by the location vendor.

As a final precaution, the administrator ran the System Security configuration utility, which identified additional steps they could take to further secure systems. These steps include disabling the primary portal administrator account, creating an automated backup schedule, and importing appropriate certificates to administrative end points to ensure the system is fully secured.

Tutorial 4: Import certificates into the ArcGIS Enterprise administrative end points

In this chapter, we discussed the importance of importing certificates into ArcGIS Enterprise end points. There are three locations where an administrator will need to import CA-signed or domain-signed certificates to fully secure administrative-level communication between the ArcGIS Enterprise portal, ArcGIS Sever, and ArcGIS Data Store.

Portal for ArcGIS

1. In a browser, sign in to your ArcGIS Enterprise portal organization site as an administrator.

2. From the **home** end point, navigate to the **portaladmin** end point.

   Tip: For example, https://gis.example.com/portal/home will now be https://gis.example.com/portal/portaladmin.

   The page presents the ArcGIS Portal Administrator Site Directory.

3. Click **Login** and sign in with your administrator credentials.
   - What important information do you see on this screen?

4. From **Resources**, select **Security** and click **SSLCertificates**.

5. Select **SSL Certificates**.

   This page contains information about all the imported certificates in ArcGIS Enterprise. By default, a self-signed certificate **portal** is present to run the system. In secure configurations, this certificate will be changed.

6. Note the certificate listed as the **Web Server SSL Certificate**.

   When configured, the domain-signed or CA-signed certificate will appear here. If **portal** is the listed certificate, your environment is still using the self-signed certificate, which will need to be replaced.

7. If **portal** is listed, contact your IT department to acquire either a domain-signed or CA-signed certificate. Additionally, acquire a root certificate for your organization.

    If a nonstandard certificate is listed, it is likely that another certificate was expected. You can investigate this certificate by clicking on it from the drop list to learn more.

    The Web SSL Certificate is a private certificate used to establish trusted communication between a client and a server. Since ArcGIS Enterprise can play both the role of a client and a server, it is helpful to import both public and private certificates.

8. From **Supported Operations**, click **Import Existing Server Certificate**.

9. Enter the necessary details, which include a **Certificate password**, an **Alias**, and the file path to the private certificate.

> *Tip: If the certificate does not contain a password, it is not a private certificate. Work with your IT team to solve this issue.*

10. Click **Import** to add the private certificate to the certificate list.

11. Back on the **SSL Certificates** page, click **Import Root or Intermediate Certificate**.

12. Provide the details for the certificate including the **Alias** and the certificate path.

    You may optionally choose not to restart Portal for ArcGIS; however, the root certificate will not take effect until a restart of the service is performed.

13. Click **Import**.

Import Root or Intermediate Certificate

> **Warning**
>
> Unless selected otherwise, executing this operation will automatically restart the portal.
>
> This is necessary for the changes to take effect.
>
> This restart will take a couple minutes to complete and cause your portal resources to be temporarily unavailable.
>
> To verify that the restart has completed, log in to the Portal Administrator Directory again before continuing.

Alias: [   ]

Do not restart the portal after import: ☐

File *   [Choose File] No file chosen

Format   [HTML ▾]

[Import]

Supported Interfaces: REST

Back on the **SSL Certificates** page, note that the active Web Server SSL Certificate is still listed as **portal**. You will need to set the fresh certificate as the Web Server SSL Certificate.

14. From **Supported Operations**, click **Update**.

15. Change the **Web Server SSL Certificate** to the alias that you imported in the previous steps.

16. Click **Update** to restart the Portal for ArcGIS service.

With the addition of this certificate, you have ensured that administrative connections in the ArcGIS Enterprise portal are secure.

ArcGIS Server

The process for updating the ArcGIS Server certificate is similar to the process for updating the ArcGIS Enterprise portal but with key differences. Because ArcGIS Sever may support multiple machines on one site, each machine needs to have its own certificate.

1. Return to the **home** end point of the ArcGIS Enterprise portal organization site. Using the **Service URL**, navigate to the **admin** end point.

> *Tip: To find the Service URL, navigate to Organization > Settings > Servers > Service URL.*

2. Because this ArcGIS Server site has been federated, you can sign in with a token from the ArcGIS Enterprise portal or by using the Primary Site Administrator (PSA) login.
    *   To get a token from the ArcGIS Enterprise portal, click the listed link and enter administrator-level credentials to generate a token for use with ArcGIS Server.
    *   The PSA login is the initial account used to create the ArcGIS Server site before federation.

3. From **Resources**, click **Machines**.

    The machines listed on the Registered Machines page make up the ArcGIS Server site.

4. Click the first machine.

    This page contains information about the machine's health, as well as the SSL certificate being used.

5. From **Resources**, click **sslcertificates**.

    Unlike the ArcGIS Enterprise portal, the default alias for the self-signed certificate included in the ArcGIS Server installation is listed as **server**.

    The **SSL Certificates** page—and the process for importing certificates from this point forward— matches that of the ArcGIS Enterprise portal. To import a server certificate, refer to the previous Portal for ArcGIS section.

ArcGIS Data Store

Unlike the ArcGIS Enterprise portal or ArcGIS Server, Data Store does not have an administrative page through which SSL certificates may be updated. Administrators must run a command line utility to replace the SSL certificate.

1. Access the ArcGIS Data Store machine (these steps may vary slightly between Windows and Linux).

2. On the machine, go to the ArcGIS Data Store program files. In the **tools** directory, locate and run the **replacesslcertificate** utility.
   - To replace the certificate used for communication web server, run the tool with the webserver option.
   - To replace the certificate used for communication among other datastore machines, run the tool with the appropriate ArcGIS Data Store type option.

   More details on the **replacesslcertificate** utility may be found in the following product documentation, "Replace ArcGIS Data Store SSL Certificates" at links.esri.com/GTKEnterprise-SSLCertificate.

Verify your system security configuration

The Esri Software and Security Team has created a set of tools and applications that can be configured to run on your ArcGIS Enterprise deployment. The content of these tools is based on the ArcGIS Enterprise Hardening Guide.

1. Navigate to the ArcGIS Trust Center at **trust.arcgis.com**.

2. At the top right of the screen, click **Launch Security Adviser**.

   The ArcGIS Security and Privacy Adviser page opens.

3. Enter the **Portal URL** for your ArcGIS Enterprise organization. Ensure that it has a **portal** end point.

4. On a separate tab, navigate to your ArcGIS Enterprise organization site and sign in as an administrator.

5. On the **Content** tab, click **New item**.

6. In the **New item** window, select **Application**. Then apply the following settings:
   - **Application type**: Web mapping
   - **URL**: https://desb9z4fk5fbl.cloudfront.net

7. Click **Next** and save the application with your own details.

8. On the app's item page, click the **Settings** tab. Scroll down to the **Credentials** section and click **Register application**.

9. For **Redirect URLs**, type the same URL you used to create the app.

10. Click **Register**.

Your application is now registered. A **Client ID**, **Client Secret**, and **Temporary Token** are generated to access the application.

11. Copy the **Client ID**. Return to the ArcGIS Security and Privacy Adviser tab. Paste the **Client ID** in the **App ID** field.

12. Click **Go To Enterprise Login**.

Running this tool will display any security issues that need to be addressed or considered, as defined by the ArcGIS Enterprise Hardening Guide.

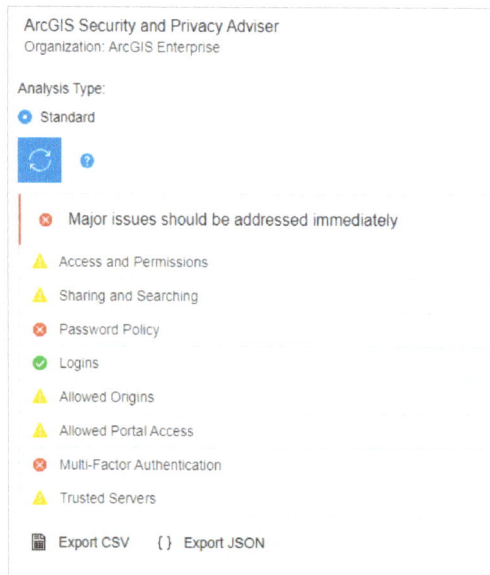

Scanning ArcGIS Server

ArcGIS Server includes a Python script tool, serverScan.py. This tool identifies security configuration options and adds a severity figure. You will run the tool to scan the ArcGIS Server machine.

1. Access the ArcGIS Server machine.

2. Navigate to the location of the ArcGIS Server installation. Open **tools** and then open **admin**.

3. Run the **serverScan.py** tool.

    This tool will require the PSA credentials used to configure the ArcGIS Server site at installation.

    This tool will also require a directory to deposit the security scan.

    An example input (for Windows deployments) may look like the following:

    ```
    python serverScan.py -n gisserver.domain.com -u admin -p
    my.password -o C:\Temp
    ```

4. Observe the outputs of the tool.
    - What possible changes may you need to make to improve security?

Scanning the ArcGIS Enterprise portal

Similar to ArcGIS Server, Portal for ArcGIS includes a Python script tool: portalScan.py.

1. Access the machine that is running Portal for ArcGIS.

2. Navigate to the location of the Portal for ArcGIS installation. Open **tools** and then open **security**.

3. Run the **portalScan.py** tool.

    This tool will require administrator-level credentials. This tool will also require a directory to deposit the security scan output.

An example input (for Windows deployments) may look like the following:

```
portalScan -n portal.domain.com -u admin -p my.password -o C:\
Temp
```

4. Observe the outputs of the tool.
    - What possible changes may you need to make to improve security?

Take the next step

If you have questions about whether a particular security configuration setting is appropriate, bring the scan results to the people responsible for information security in your organization and further discuss any additional steps you should take to harden your ArcGIS Enterprise deployment.

Summary

In this chapter, we introduced the three levels of security that are relevant to ArcGIS Enterprise: ArcGIS Enterprise identity-based security, software-level security, and network-level security. The goal of a good security strategy is to find an intersection between function and access. To reach this goal, ArcGIS Enterprise administrators must work closely with their IT counterparts to identify the access requirements of features that are hosted on the internet and the privacy requirements of the data in ArcGIS Enterprise.

CHAPTER 5

Going beyond the ArcGIS Enterprise base deployment

Objectives

- Review distributed computing principles and appropriate scaling options in ArcGIS Enterprise.
- Explore options to extend ArcGIS Server.
- Discuss ArcGIS Enterprise on Kubernetes.

Introduction

In the previous chapters, we discussed ArcGIS Enterprise in its most essential form—a base deployment. As business needs change, administrators need to be prepared to scale their deployments of ArcGIS Enterprise to meet growing demands. The type of scaling needed depends on the type of work to be completed or performance limitation to be solved. ArcGIS Enterprise supports scaling to distribute workflows and add functionality through ArcGIS Server roles and extensions.

In this chapter, we will focus on what options administrators have when considering how to scale or extend their ArcGIS Enterprise deployments. We will review the basics of work-load separation and how to apply different methods to scale out environments that support various workflows. We will also explore the ArcGIS Server roles and extensions, defining their use cases as well as some best practices for deploying these roles.

The chapter will conclude with a review of ArcGIS Enterprise on Kubernetes. Using containerized architecture, ArcGIS Enterprise on Kubernetes seeks to improve system responsiveness and availability through automated scaling. Many of the concepts discussed in this book regarding architecture considerations do not apply as clearly to Kubernetes architecture, but it is important to understand which needs ArcGIS Enterprise on Kubernetes has been designed to address.

Distributed GIS and workload separation

As we mentioned in chapter 2, ArcGIS Enterprise is made up of four primary components: ArcGIS Enterprise portal, ArcGIS Server, ArcGIS Data Store, and ArcGIS Web Adaptor. The base deployment of ArcGIS Enterprise requires each of these components to be deployed and configured to form a functional deployment. However, this is the minimum viable product of ArcGIS Enterprise, so what if we needed to push it further?

Recall the core concepts of a distributed GIS: Client software is used to offload expensive workloads to servers to improve data efficiency and process flow. The client software does not always have to be ArcGIS Pro or other desktop offerings. Depending on the work being performed and the location of the service, the ArcGIS Enterprise portal may also be a client, a user's browser, or ArcGIS Server itself. In these cases, understanding the origin of the work and the types of resources needed to complete a job becomes important from a scaling perspective.

You can consider scaling in ArcGIS Enterprise in two ways:

- **Scaling up**: Add resources to your existing machines (RAM, CPU, hard disk space). Scaling up can be beneficial if many services exist on a site.
- **Scaling out**: Add more machines to sites/stores. Scaling out may make sense if user demand grows or to enable high availability.

We will cover how these options apply to ArcGIS Server, Portal for ArcGIS, and ArcGIS Data Store later in this chapter.

Administrators need to make informed decisions on how to scale or grow ArcGIS Enterprise. Because of the resources needed to be successful, scaling any enterprise system introduces added levels of complexity, upkeep, and cost. Thoughtfully analyzing and balancing system requirements against investment in new infrastructure will promote sustainable growth of your organization's capacity.

ArcGIS Server

An ArcGIS Server site is a single configuration of ArcGIS Server that fulfills a specific role or process. An ArcGIS Server site can be configured to run on either a single machine or a multimachine configuration across multiple virtual machines, and the distinction between a site and the machine or machines the site comprises is key to the overall understanding of ArcGIS Server scalability. One common example of a server site is the hosting site that is configured in an ArcGIS Enterprise base deployment. As your user base matures and more projects get introduced, it may be best to scale out by adding more machines to the hosting site. The primary benefit of scaling out is providing more resources to a site without needing to rely on adding resources to a single machine. This introduces some redundancy and safety to scaling the system out.

When scaling out an ArcGIS Server site, specific requirements must be met to ensure a successful implementation. One key area is the storage location of the configuration store; all too often, multimachine sites will have their configuration store located on one of the machines in the active site, and if that machine should go down for any reason, you stand to lose access to the site completely. Instead, ensure that the files in the configuration store are available to each machine in a site regardless of the hardware status of the individual machines in the site—this is critical to the long-term success of a multimachine site. There are also version-specific considerations, so make sure you consult the ArcGIS Server documentation for your version.

But what if a lot of services are running on ArcGIS Server? This is where scaling up an ArcGIS Server site makes the most sense. Let's consider an ArcGIS Enterprise deployment that has a dedicated ArcGIS Image Server. Its capabilities include servicing referenced image services, enabling raster analysis in the ArcGIS Enterprise portal, and storing hosted image services. If you find that users are running more raster processing over time and the performance of those tools is degrading, it may be a good idea to scale up the ArcGIS Image Server site to include more resources.

We will discuss ArcGIS Server roles later in the chapter; however, the key takeaway of this section is to understand when to scale a site out versus when to scale a site up. Scaling ArcGIS Server in either direction will generally increase the efficiency of services or the total user access capacity to the site. However, it is important to realize that you need to partner with your database administrators and IT specialists to scale other dependencies appropriately, such as databases and network connections.

Portal for ArcGIS

The scale-up concepts for the Portal for ArcGIS remain consistent with ArcGIS Server. If you find that more users are accessing ArcGIS Enterprise to manage their content or to create apps, adding more resources, including RAM, CPU, and hard drive space, is a good way to handle this load.

The key difference comes when we talk about scaling out Portal for ArcGIS. Unlike ArcGIS Server, you cannot add more than two machines to a single deployment of Portal for ArcGIS. Additionally, adding a second machine to the ArcGIS Enterprise portal does not achieve the same performance or scalability objectives as with ArcGIS Server. In the context of Portal for ArcGIS, scaling out to a second machine is the beginning of setting up a highly available, or HA, deployment.

Deploying and maintaining a successful HA deployment of ArcGIS Enterprise requires a lot of thought and intention when designing the system. To learn more about what you must consider when deploying these types of systems, access the high availability section of the ArcGIS Architecture Center at links.esri.com/GTKEnterprise-HighAvail.

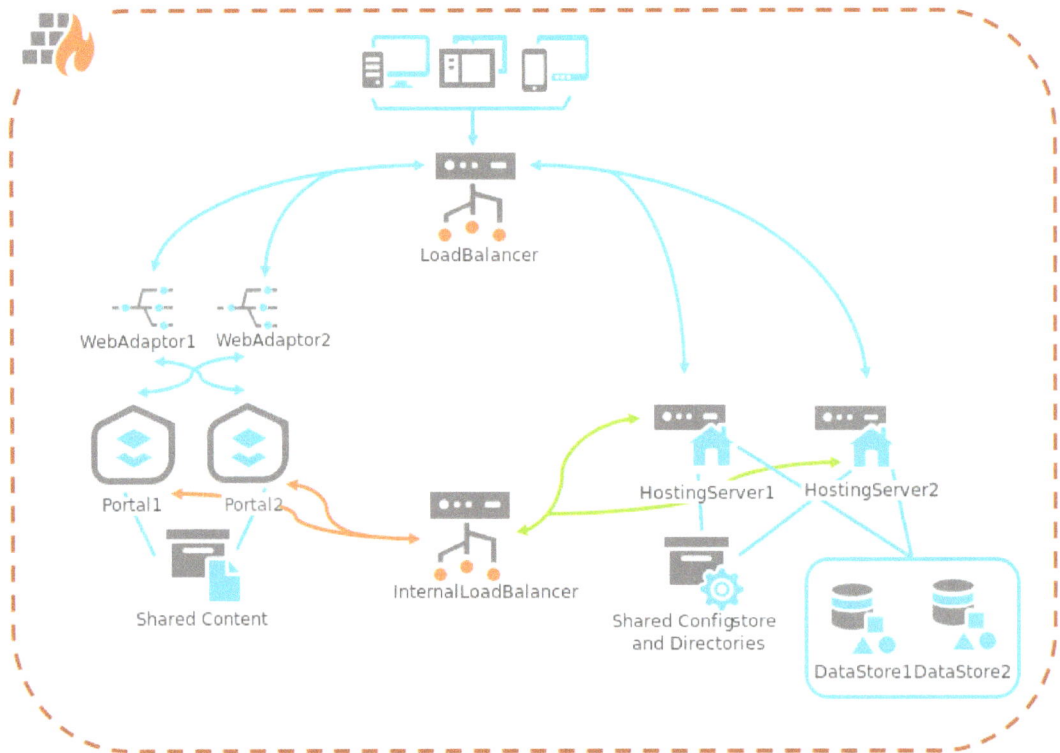

Figure 5.1. A sample of HA ArcGIS Enterprise deployment.

Unlike a multimachine ArcGIS Server site, where traffic and requests are actively serviced by all machines within a site, the ArcGIS Enterprise portal works differently. Both machines in the site will actively handle traffic simultaneously to maintain a consistency of high availability. However, only one of the portal machines will contain the active underlying database used by the Portal for ArcGIS to manage user activity. When configured correctly, the HA system will be able to switch to the secondary system.

Deploying a true HA ArcGIS Enterprise organization requires a significant amount of architectural planning and system redundancy to achieve. Redundancy can be achieved at every component of ArcGIS Enterprise by configuring separate load balancers between Portal for ArcGIS, ArcGIS Server, and ArcGIS Data Store. Each of these components may be configured to have standby deployments (as seen in the figure) referencing the same configuration stores. Although the sample provided is one solution, there are many ways to attain HA, depending on your chosen ArcGIS Server extensions and multisite deployments.

ArcGIS Data Store

Before considering scaling any type of ArcGIS Data Store, it is best to start with a fully separated instance of ArcGIS Data Store hosted on independent resources. Although it is possible to run multiple components on a single machine, operating at scale may resolve itself in resource competition between ArcGIS Enterprise components, which may negate any resources added to the machine. ArcGIS Data Store can consume a large number of resources quickly, and some ArcGIS Data Store types are configured to use a constant number of resources upon start-up for internal efficiency. The scaling rules for ArcGIS Data Store depend on the type of data store being used. At the 11.4 release of ArcGIS Enterprise, four data store types exist within the system:

- **Relational data store:** The relational data store is used to configure the base deployment of ArcGIS Enterprise. It follows the same scaling logic that Portal for ArcGIS follows. Scaling out the relational data store can be done only to configure an HA environment. This means that a second machine may be configured for a relational data store site, and this machine may be used only as a passive backup, not for additional resources. Scaling up the relational datastore machine by adding more resources, such as RAM, CPU, and hard drive space, is supported.
- **Spatiotemporal big data store:** The spatiotemporal big data store supports scaling out its resources by adding additional nodes (machines) to the instance. Because of the baseline architecture of the site, administrators must always have an odd number of nodes available. This means that if you need to scale a spatiotemporal big data store instance that has eight CPU cores and 32 GB or RAM, you will need to add at least **two more** machines of the exact resource allocation to the site at the same time for proper operation. This would total three nodes.
- **Graph store:** The graph store follows the same scaling rules as the relational data store. It supports scaling out only in the context of a passive/active HA configuration. You would add resources directly to the machine running the graph store when required.
- **Object store:** The object store supports scaling through clustering in a similar way to the spatiotemporal big data store . If an odd number of clusters are available, and resources match each one, the object store will be able to use these resources. Scaling up through adding resources to each node is also supported.

So far, we've covered different options to scale up ArcGIS Enterprise by adding CPU, RAM, and disk space to machines within the organization. We've also covered how to scale out ArcGIS Enterprise by adding machines and nodes to specific components that support these additions. But what if you wanted to add to the features that are available to your users? This is where different ArcGIS Server roles come into play.

ArcGIS Server roles and extensions

Certain features in ArcGIS Enterprise require additional federated ArcGIS Server sites licensed and configured to fulfill a specific role. At minimum, an ArcGIS Server role requires a dedicated site that has been licensed for that role. Some ArcGIS Server roles require additional underlying software of configured ArcGIS Data Store types to function.

> *Note: In addition to ArcGIS Server roles, there are also ArcGIS Server extensions. An ArcGIS Server extension is different from a role in that it enables a more focused set of tools that are used for a particular purpose. Examples of these tools include the ArcGIS Aviation Airports extension and ArcGIS Aviation Charting extension, which are used to aid in aviation-based GIS analysis. Because of the changing nature of extension availability and license requirements, we recommend checking the ArcGIS GIS Server capabilities and extensions document appropriate for your version for more information.*

ArcGIS GIS Server

This site is the foundation of your ArcGIS Enterprise base deployment. The ArcGIS GIS Server site is configured to support two primary functions:

- Act as a hosting server that supports the ArcGIS Enterprise base deployment
- Act as a publishing server that supports publishing referenced feature services from referenced data sources, such as enterprise geodatabases

For workload separation reasons, it's a good idea to have these two responsibilities spread across two sites. One ArcGIS GIS Server site would handle all hosting role responsibilities, while the other would be responsible for all referenced feature service access. In this configuration, administrators may scale up and scale out the most pertinent resource without having to overload the other feature.

ArcGIS Image Server

The ArcGIS Image Server has three primary functions:

- Enable the publishing of referenced image services.
- Enable distributed raster analysis within ArcGIS Enterprise.

• Store hosted imagery layers that are a result of some raster analytics tools.

Raster-based workflows can be incredibly resource intensive, depending on the scale and scope of the raster feature. Land classification, watershed analysis, and digital elevation model (DEM) analysis are some examples of the types of analysis that become available with an ArcGIS Image Server role enabled in ArcGIS Enterprise. The outputs of any raster analytics tool will also produce a hosted image service in your ArcGIS Enterprise content.

If working with imagery and raster analytics represents a large amount of work in your organization, you can optionally separate raster analysis and image hosting into two Image Server sites for added workflow separation.

ArcGIS GeoEvent Server

The ArcGIS GeoEvent™ Server role supports working with live data. A GeoEvent server site is different from a traditional ArcGIS Server site in that it includes a separate GeoEvent Server manager page. This page may be used to create inputs, outputs, GeoEvent Server services, and input definitions. If your goal is to see a live "blue dot" navigation experience for mobile assets on a web map, stream services make this a reality.

GeoEvent Server is capable of processing thousands of events per second when configured correctly. Using GeoEvent in a scaled capacity, it is better to have multiple independent GeoEvent Server sites processing events rather than a single multimachine site. This creates logical workload separation between inputs and outputs and reduces overall strain on the data stores handling data ingress.

GeoEvent Server can work with all supported ArcGIS Data Store types through ArcGIS Enterprise. However, if users of GeoEvent Server anticipate processing many tasks at the same time, the spatiotemporal big data store is designed to keep up with the load. In a scenario in which stream services are required, a federated GeoEvent Server is required, along with the spatiotemporal big data store configured with ArcGIS Enterprise.

ArcGIS Notebook Server

The ArcGIS Notebook Server role allows users to use ArcGIS Notebooks directly in ArcGIS Enterprise. ArcGIS Notebooks uses ArcGIS Notebook Server to perform spatial analysis and administer ArcGIS Notebook Server. ArcGIS Enterprise users who have the proper license may be able to instantiate their own notebooks directly in the browser, creating their own Python code independent of other ArcGIS Notebooks users.

From an administrative perspective, ArcGIS Notebooks allows advanced users to administer many aspects of ArcGIS Enterprise through a notebook. Setting an informational banner, kicking off backup activities, and doing anything available in ArcGIS API for Python are available for use in ArcGIS Notebooks.

63

Building on this point, ArcGIS Notebooks can be published as web tools in ArcGIS Enterprise. These web tools may be used directly through Map Viewer as a custom web tool. This means that ArcGIS Notebook Server users may publish web tools and have other users in the organization use them for their work.

ArcGIS Workflow Manager Server

The ArcGIS Workflow Manager Server role introduces a way to create guided workflows within ArcGIS Enterprise. From a quality control and quality assurance perspective, Workflow Manager can control most aspects of data entry, verification, and distribution within an organization.

Workflow Manager users can create custom workflows in the context of ArcGIS Enterprise through the Workflow Manager web app, in ArcGIS Pro, and a set of semiautomated and fully automated workflows to aid in various tasks.

ArcGIS Knowledge Server

Knowledge graphs are a graph-structured data model that can organize and represent information between entities and their relationships to each other. Through ArcGIS Knowledge Server, users can understand the context of knowledge graphs alongside geographic information.

Knowledge Server is made up of two components—an ArcGIS Server site using the Knowledge Server role and an ArcGIS Data Store type configured as a graph store. Once Knowledge Server is configured to ArcGIS Enterprise, users can analyze knowledge graph layers to understand the relationships between entities stored within the graph store.

ArcGIS Video Server

Customers who create drone imagery or any other source of geospatially tied video files can use ArcGIS Video Server to publish and stream video layers in ArcGIS Enterprise. Along with keeping the geospatial and temporal context of the video, Video Server allows users to index and search the video content through the familiar interface of ArcGIS Enterprise.

Video Server requires two components—an ArcGIS Server site licensed with the Video Server role and ArcGIS Data Store configured with an object store. Because of the potential size of video collections, system administrators should ensure that ample resources are available for video file storage.

Now that we've discussed the roles of ArcGIS Server, there are some best practices for system deployment and architecture. In an ideal deployment, all server roles should be deployed on their own machines. Supporting resources, such as ArcGIS Data Store types, should also be deployed on their own machines. Although planning for this type of distribution increases system complexity and upkeep tasks, it will also prevent resource competition across ArcGIS Enterprise.

ArcGIS Enterprise on Kubernetes

ArcGIS Enterprise on Kubernetes is based on powerful containerization technology. Most, if not all, architecture conversations presented in this book do not have parallels to the Kubernetes deployment option. To explain why, let's explain some key concepts.

Kubernetes is an open-source container management software. It is designed to orchestrate, deploy, and scale a set of containers within a system to flex with the growing demand of a system. In this context, a container is a piece of software that is designed to run alongside other containers to fulfill the needs of a process or task.

The key difference between software made to run on Kubernetes and Windows or Linux is in this containerization concept. Windows and Linus offerings are made up of the four primary software components: Portal for ArcGIS, ArcGIS Server, ArcGIS Data Store, and Web Adaptor. ArcGIS Enterprise on Kubernetes takes the key functions and features of these four components and breaks them up into many containers.

The benefit of containerization is that the system can scale up and out automatically when presented with load. For example, imagine a raster processing task in which a task is submitted to classify a large area of land. Through containerization, ArcGIS Enterprise on Kubernetes spins up more containers that are responsible for the classification job to meet the demand. This capability results in faster processing time and an overall more efficient deployment. The efficiency comes in the form of ArcGIS Enterprise on Kubernetes spooling down containers it no longer needs after a task is completed.

ArcGIS Enterprise on Kubernetes also benefits from a backup process. Unlike other deployments, the backup task is handled directly through the enterprise manager of ArGIS Enterprise on Kubernetes. Because the ArcGIS Server and Portal components no longer apply in the same way, many of the administrative tasks needed to run the deployment have been placed here.

ArcGIS Enterprise on Kubernetes also has the benefit of a more efficient upgrade process. Because each container gets upgraded asynchronously, the system can upgrade quickly. This translates to shorter downtime as well as a less intensive upgrade process. One relevant detail to note here is that the ArcGIS Enterprise on Kubernetes life cycle is different from Windows and Linux offerings. Because of the rapid updating cycle, as well as the constant changes introduced in Kubernetes, it is expected that the administrators will upgrade often. For more details, review the article "ArcGIS Enterprise on Kubernetes Life Cycle," at links.esri.com/GTKEnterprise-KubernetesLifeCycle.

5

From a usage perspective, you will find few differences in this technology compared with the look and feel of ArcGIS Enterprise. Content, group, and user management are consistent with Windows and Linux offerings. ArcGIS Enterprise on Kubernetes comes with familiar web apps, including ArcGIS Experience Builder and ArcGIS Instant Apps. Not all ArcGIS Server roles are included in ArcGIS Enterprise on Kubernetes; however, more roles are being included in every release. The Capabilities section of the version of ArcGIS Enterprise you are deploying will have more information from a function perspective.

A successful deployment of ArcGIS Enterprise on Kubernetes is built on a foundational grasp of Kubernetes at the architectural and networking level. If you become curious about these capabilities or require certain scaling features that Kubernetes offers, talking to your IT department about its experience with deploying Kubernetes clusters and understanding their capabilities is a vital first step toward success.

Fictional user story

Yvonne Akindele, one of the GIS Systems administrators at SuperBiz International, received a message from a user that a map service they need is responding too slowly to requests. Checking the statistics for the service, Yvonne confirmed that service response time has increased substantially in the last hour. The total number of requests coming into the ArcGIS Server site, however, had not changed from the typical level. Investigating further, Yvonne discovered that the CPU and memory usage of the machine where ArcGIS Server is installed had also increased in conjunction with the service response time degradation. Yvonne tracked the increased system resource usage to a single, large geoprocessing service job.

Because the geoprocessing service and map services were published to the same ArcGIS Server site, they contended with each other for system resources. Yvonne discussed possible solutions with her colleagues. The map service response time increase was bad enough that they could not just ignore the problem. They rejected scaling vertically by adding more CPU and RAM to the ArcGIS Server site machine because they could not accurately predict the future growth of resource usage from the geoprocessing service. Underpredicting would mean running into the same resource contention problem in the future. Overpredicting would mean paying for idle resources. They rejected the possibility of scaling horizontally by adding more machines to the ArcGIS Server site for the same reason.

In the end, they decided to federate an additional GIS Server site and republish the geoprocessing service to that site. Large geoprocessing jobs would no longer contend for resources with map services. They could independently scale the resources of the second ArcGIS Server site in the future, if necessary.

Tutorial 5: Investigate distributed computing in ArcGIS Enterprise

Your deployment of ArcGIS Enterprise may already make use of distributed computing to separate workloads or provide redundancy. In this tutorial, you will investigate how, and whether, you are already taking advantage of distributed computing.

5

Investigate the ArcGIS Enterprise portal

1. Sign in to your ArcGIS Enterprise portal organization as an administrator.

2. Change the organization site URL to point to the Portal Administrator Directory.

   Tip: For example, if your organization site URL is https://gis.example.com/portal/home, your Portal Administrator Directory is located at https://gis.example.com/portal/portaladmin.

3. On the **Site Root** page, in **Resources**, click **Machines**.
   • How many machines are listed?

   An HA ArcGIS Enterprise portal will list two registered machines. A single machine represents a single point of failure: If that machine goes down, the capabilities of your ArcGIS Enterprise portal will not be available.
   • What are the fully qualified domain names of each machine?

4. Click the **status** link for one of the machines.
   • What is the machine's status?

5. At the top of the page, click the **Home** link to return to the **Site Root**.

Investigate federated ArcGIS Server sites

1. Within **Resources**, click **Federation** and then click **Servers**.
   • How many federated servers are listed?

2. Click the link for the hosting server.
   • What is the URL for the hosting server?

3. Copy the URL for the hosting server and paste it on a new browser tab. Add **/manager** to the end of the URL and then navigate to that page to access **Server Manager**.

> *Tip: For example, if your hosting server URL is https://gis.example.com/server, navigate to https:// gis.example.com/server/manager.*

4. At the top of the page, click the **Site** tab.

5. On the left, go to the **Machines** section.
   • How many machines are listed for this server site?

   Multiple listed machines indicate an HA server site with an active-active configuration. Each machine is capable of handling requests for any service published to the server site, increasing both the capacity and resilience of the site.
   • What are the fully qualified domain names for each machine?
   • Are the fully qualified domain names the same as any machine used for the ArcGIS Enterprise portal?

   If multiple components are running on the same machine, they will contend for system resources. Problems in one component are more likely to negatively affect other components if they are running on the same machine.

6. Repeat the previous steps for any other federated servers listed in the Portal Administrator Directory.

Investigate ArcGIS-managed Data Store items

1. Change the URL to point to the ArcGIS Server Administrator Directory.

> *Tip: For example, if your Server Manager URL is https://gis.example.com/server/manager, your ArcGIS Server Administrator Directory is located at https://gis.example.com/server/admin.*

2. Sign in to the ArcGIS Server Administrator Directory using the credentials for the Primary Site Administrator or by following the directions to acquire a portal token.

3. From **Resources**, click **data**. Then click **items** and go to the **/enterpriseDatabases** link.

4. Click the link for the **Child Item** that has **AGSDataStore** in its name.

   This is the relational data store that holds data for hosted feature layers.

5. Below **Data Item Properties**, click **machines**.
   • How many machines are listed?

   Multiple machines indicate that there is redundancy at the relational data store level. If the primary machine fails, the standby will be promoted to primary.
   • What are the fully qualified domain names of each machine?
   • Are any of the machine names the same as those used for your ArcGIS Enterprise portal or any ArcGIS Server site?

6. Click the name of one of the machines.
   • What role does this machine have?

7. At the top of the page, click the **items** link to return to the list of **Root Data Items**.

8. Click the **/nosqlDatabases** link.

   The child items on this page have information about the other ArcGIS-managed data stores registered with the hosting server, such as the object store or spatiotemporal big data store.

9. Repeat the previous steps for each of the child items listed.

10. Return to the **Site Root** of the Portal Administrator Directory on your previous browser tab.

Investigate the Portal for ArcGIS Web Adaptor components

1. From the **Site Root** of the Portal Administrator Directory, within **Resources**, click **System** and then click **Web Adaptors**.
   • How many web adaptor components are listed?

Multiple web adaptors indicate resiliency at the web adaptor tier. If one of the Web Adaptor components fails, requests can still be routed to ArcGIS Enterprise.

2. Click the name of one of the **Web Adaptor** components.
   - What is the IP address of the Web Adaptor component?

Summary

In this chapter, we covered the basic concepts on how to extend your deployment of ArcGIS Enterprise to meet new requirements. Extending ArcGIS Enterprise can include scaling resources by adding hardware to an existing deployment to meet new volume requirements. Adding new server roles to an ArcGIS Enterprise organization will add functionality to an existing site. Lastly, we touched on ArcGIS Enterprise on Kubernetes and how it can scale to match volatile demand for services.

PART 2

Administering ArcGIS Enterprise

THE NEXT FIVE CHAPTERS SERVE AS A PRIMER TO SUCCESSFULLY manage and administer ArcGIS Enterprise. The topics covered in these chapters include getting familiar with the organization's site, managing users, managing content, and publishing data. Administering ArcGIS Enterprise builds on a successful deployment and implementation of ArcGIS Enterprise as defined in part 1. However, as your organization matures and gains more users, it becomes important to effectively manage users and their content. As with the previous section, the topics and ideas introduced here are not prescriptive, meaning you should modify them to fit your organization's specific needs and goals. By following the concepts introduced in these chapters, you will be able to effectively administer an ArcGIS Enterprise organization.

CHAPTER 6

Getting familiar with the ArcGIS Enterprise portal

Objectives

- Navigate the ArcGIS Enterprise portal.
- Identify administrative tasks performed in the ArcGIS Enterprise portal.
- Determine whether your ArcGIS Enterprise portal configuration meets your needs.

Introduction

In this chapter, you'll familiarize yourself with the ArcGIS Enterprise portal and its website, which you can use to access the geospatial capabilities enabled by the infrastructure of your ArcGIS Enterprise deployment. If you are familiar with the ArcGIS Online interface, the website for ArcGIS Enterprise looks nearly identical.

Exploring the ArcGIS Enterprise portal

Put simply, the ArcGIS Enterprise portal makes GIS accessible for users of all experience levels. It provides geographic viewers for those just starting out, while experienced users can connect to the portal from advanced applications and use its content in analysis and mapping workflows.

Navigation links

The ArcGIS Enterprise portal is composed of several website pages, accessed from navigation links at the top of the portal. Administrators can configure page visibility for members, but several pages are available by default.

- The **home page** is for branding your ArcGIS Enterprise portal. You can configure this page to provide a custom name, images, logo, and other information to provide a friendly landing page. By default, everyone who accesses your portal will see the home page first.

- The **gallery page** is a collection of easily accessible content items of your choosing. Generally, these will be items shared publicly or across your entire organization and have broad applicability to your users. By default, the Gallery shows all the items in your organization, which may not be appropriate for organizations with a lot of content. You can configure the content in the Gallery by selecting a specific group of items instead. As you configure Gallery content, it is important to strike a balance between including enough content items to be useful without including so many items that it becomes overwhelming and difficult to search through. To use the Gallery effectively, you should actively keep it updated by adding and removing content items from the Gallery as their relevance changes.

- The **map page** gives users access to the Map Viewer application. Map Viewer allows users to add data layers; configure layer symbology, filters, labels, and pop-ups; perform spatial analysis; and create interactive web maps.

- The **scene page** provides access to the Scene Viewer application. Scene Viewer is like Map Viewer, except that it enables users to work with 3D scenes instead of 2D maps.

- The **groups page** lists the groups a user has permission to see. Only organization members can belong to groups. Groups organize members and content to make it easier to ensure that every member has access to all the content they need and none that they don't. From the groups page, a member can create groups, join groups, accept invitations to groups, see the content of groups they belong to, and configure settings for groups they own or manage.

- The **content page** lists all the content the user has permission to see. Only members can own content. This content can be layers, maps, apps, files, and more. From the content page, a member can access the item page for each item, control the sharing of their items, add or delete items, and organize the items into folders.

- The **organization page** provides information on the ArcGIS Enterprise organization itself. Only members can access this page. Members who do not have administrative privileges in the organization will see only limited information on this page. Administrators use this page to perform a variety of tasks to manage the organization, such as adding or removing members, managing licenses, configuring security options, and federating ArcGIS Server sites.

Each member can configure for themselves the page that opens when they first sign in. Administrators can configure which pages are visible to certain members to control their access to the capabilities of the ArcGIS Enterprise portal.

Anatomy of an organization's URL

The installation and configuration of the various components of ArcGIS Enterprise will affect the URL of the ArcGIS Enterprise portal. Most users will access the organization at a URL that looks something like this:

https://webserver.example.com/portal/home

| Scheme | Host | Context | App |

Figure 6.1. Example of an ArcGIS Enterprise portal URL.

- **https://** refers to the scheme that is used to transfer information to and from whichever application the user is using to access the portal. The https scheme is encrypted to prevent potentially malicious actors from intercepting traffic. It is possible to configure ArcGIS Enterprise to allow the unencrypted http protocol in addition to https, but it is a security best practice to disable http access.
- **webserver.example.com** is the fully qualified domain name (FQDN) or DNS alias of the web server machine where the ArcGIS Web Adaptor is installed. It is a best practice to have users access the portal through either the web adaptor or a third-party load balancer/reverse proxy. That way, users communicate directly only with the web server and do not send requests to the Portal for ArcGIS service.
- **portal** is the name you specified when you installed the web adaptor or third-party load balancer and configured it with your ArcGIS Enterprise portal. This name is referred to as the portal's context.
- **home** is the specific web app the user is using to access the portal. Unlike the other parts of the URL, this name is not user configurable.

As an administrator, you should be aware of other URLs that may be useful for managing and troubleshooting your ArcGIS Enterprise deployment.

https://portalmachine.example.com:7443/arcgis/home

| Scheme | Host | Port | Context | App |

Figure 6.2. Example portal URL when accessed directly.

With this URL, you can bypass the web adaptor or third-party load balancer and send requests directly to the Portal for ArcGIS service.

- **portalmachine.example.com** is the FQDN or DNS alias of the machine where Portal for ArcGIS is installed. In a single-machine deployment where the web adaptor is installed on the same machine as Portal for ArcGIS, the FQDN will be the same.
- **7443** is the port number that this service is listening on. Because it uses a port other than the default https port of 443, you must specify the port number in the URL. This port number is not user configurable. The Portal for ArcGIS service always listens on port 7443.
- **arcgis** is the context when you access the portal directly because there is no web adaptor to configure a custom context. This value will always be arcgis. Users cannot configure it.

https://webserver.example.com/portal/portaladmin

| Scheme | Host | Context | App |

https://portalmachine.example.com:7443/arcgis/portaladmin

| Scheme | Host | Port | Context | App |

Figure 6.3. Example of Portal Administrator Directory URLs when accessed through Web Adaptor (*top*) and directly (*bottom*).

These URLs take you to the Portal Administrator Directory. This web app exposes additional administrative functionality. For example, you can use the directory to query the portal logs or update the transport layer security (TLS) certificates.

Member profile and settings

If you are currently signed in, the portal will show the full name and username associated with your member account. Once signed in, you can access your profile, access your settings, and sign out.

Your profile contains information about you. Your profile picture, name, and bio provide personal information. Your item gallery is a configurable set of your shared content items that you want users to see when they access your profile. Your profile also lists the publicly discoverable groups you belong to.

By default, your member profile is not private because its main purpose is to be your public-facing information page. Users can reference your public profile to help access and understand the context of your content items. You can, however, configure your profile to be private.

Fictional user story

Linda Jackson, executive director of the Becken Pond Conservation Society, is a lawyer by training and does not have a GIS background. She has a Viewer user type and Viewer role in the ArcGIS Enterprise portal so that she can view finished maps and share them with partners. Linda has a bookmark in her browser for https://maps.bpcs.org/portal/home that takes her to the organization's ArcGIS Enterprise portal. But today, when Linda clicked the bookmark, she got an HTTPS Error 404 message, indicating that the portal was not available.

Linda contacted Jim Yazzie, the BPCS GIS expert, for help. Jim first tried to access the portal logs by navigating to the Portal Administrator Directory at https://maps.bpcs.org/portal/portaladmin. He received the same error Linda got, which informed him that the problem was not local to Linda's computer. Jim's first suspicion was that the problem was related to the web adaptor.

The Portal for ArcGIS Web Adaptor is installed on one virtual machine (VM), whereas the other ArcGIS Enterprise components are all installed on another single VM with an FQDN of gis.bpcs. org. To check the logs, Jim bypassed the web adaptor by navigating to https://gis.bpcs.org:7443/ arcgis/portaladmin and accessing the Portal Administrator Directory, as expected. Querying logs did not find any errors, which was further evidence of a web adaptor problem. Jim then went to https://gis.bpcs.org:7443/arcgis/home and was able to access the ArcGIS Enterprise portal, confirming his suspicions.

Investigating further, Jim saw that the VM where the web adaptor is installed was not currently running and would not start because there was not enough RAM available on the host. Jim reconfigured some of the other VMs on the host to free up enough memory and successfully restarted the maps.bpcs.org web server. That work resolved Linda's issue.

Your settings determine how your account interacts with the capabilities of ArcGIS Enterprise, such as determining the correct formatting for dates or numbers. Your settings also include nonpublic information, such as your email address and the ArcGIS capabilities you can access because of your user type, role, and add-on licenses. If you are using a built-in account, you can change your password and security question from your settings.

With administrative privileges, you can change your profile and the settings of other members. You may want to use the capability to provide standard defaults, such as adding headshots to member profile pictures. For members with built-in accounts, administrators can also reset those members' passwords if they are forgotten or compromised.

Fictional user story

Frida Gomes is one of the environmental scientists who volunteers for the Becken Pond Conservation Society. Since college, people have generally referred to Frida by her nickname, "Go-Go," even in her professional capacity as a field researcher. Her name and email address in the ArcGIS Enterprise portal for BPCS reflect that nickname.

Frida has always been ambivalent about being called Go-Go, and a recent promotion in her job has prompted her to adopt a different professional persona. She signed in to the BPCS portal and navigated to her profile. She was able to edit her own name easily, but she didn't see a way to edit her email address from her profile.

Frida contacted Jim Yazzie, the BPCS GIS expert, for help. Jim said that her email address isn't part of her public profile. The only other people who can see it are Jim, the backup administrator, and the service provider who manages the BPCS email server. Jim explained that Frida does not have the appropriate administrative privileges to change her own email address, but he would change it for her. Because Jim has configured email settings for the organization, email notifications sent by the system will automatically go to Frida's new email address.

Configure the organization home page

Because the home tab is the default landing page for everyone who reaches the portal, you will likely want it to reflect your organization's branding. Starting with ArcGIS Enterprise 10.9, you can do that using the component-based home page editor if you have the appropriate administrator privileges. You can access the editor app from the home page by clicking the **Edit** home page option. You can access the same editor from the settings on the organization page. The editor allows you to configure several aspects of your home page.

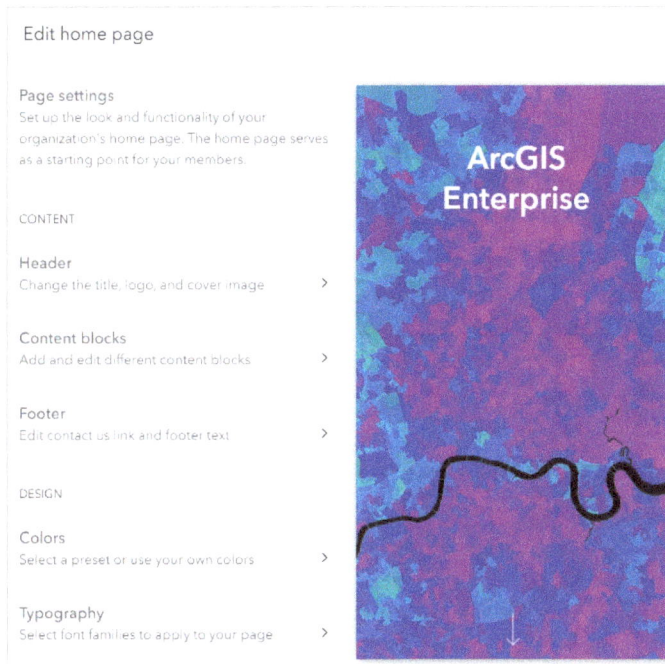

Figure 6.4. Example of the default home page.

Editing the home page:

- The header appears at the top of the home page. You can change the name of your organization from the default "ArcGIS Enterprise." You can also upload a logo, change the background image, and configure the layout of all the elements in the header.
- You can add content blocks below the header to highlight group content in an item gallery, add text, or add links to resources.
- The footer appears at the bottom of the home page. You can change the color and add text to the footer.
- You can configure the color theme, which affects the colors that are available in the other sections. You can pick from several preset themes or create your own custom theme to match your organization's branding colors.
- The typography of the home page affects the font families that the home page uses for text. As with colors, you can pick a preset theme or create a custom theme.

Fictional user story

Jim Yazzie, the Becken Pond Conservation Society GIS expert, has never been too concerned about the look of the organization home page because it was only used internally. As a result, he never changed it from the default.

But now the BPCS would like to publicly highlight some of its work to raise the organization's profile and improve fundraising.

Jim is not a web designer, and he wasn't sure he was the right person to take on the task of ensuring that the organization home page reflected BPCS branding. Alis Bourne, the BPCS communications director, had the relevant design skill set but did not have the appropriate permissions in the ArcGIS Enterprise portal to edit the home page.

Giving Alis a default administrator role would not have been appropriate. Alis should not be authorized to make changes to services, manage members, or nearly any of the other portal tasks that administrators can do. Instead, Jim decided to create a custom limited administrative role for Alis with just the privilege to manage the portal's home page settings.

Alis used the home page editor to update the page. None of the available preset color schemes matched the BPCS brand colors, so they created a custom scheme and applied it to the header and footer. Alis chose Noto Serif for titles and Roboto for body text to match the BPCS official communication typefaces. They uploaded an image of the official BPCS logo and chose a popular picture of the wetlands to serve as a backdrop. Alis added their standard call to action link to the footer to facilitate BPCS fundraising efforts.

Alis asked Jim to add the top four most impactful content items to the organization's Featured Maps and Apps group and added that group as an item gallery content block. The existing thumbnails for those items were low resolution, so Alis and Jim worked together to improve the look of the thumbnails for display in the content block.

After the customization, the home page reflects the image of the organization that the BPCS wants to project in public.

Maintenance tasks

The ArcGIS Enterprise portal gives administrators tools for performing a wide variety of maintenance tasks for an ArcGIS Enterprise deployment. You will explore the workflows involved in performing these tasks more in-depth in later chapters. The administrative tools are largely located on the portal's organization page and separated into a few categories.

- **Manage members**: Tools for managing members include searching, adding, and deleting members. Administrators can change a member's add-on licenses, user type, and role. Administrators can also manage a member's content items.
- **Manage licenses:** You can see at a glance as administrator how many licenses are available for each user type and each add-on license. As an administrator, you can also

import a new Portal for ArcGIS license file to change the user types, add-ons, or extensions you have available or renew your licensing. Newly imported files replace your existing user types and add-on licenses. Make sure the file includes all the licenses your organization needs.

- **View organization status:** The portal includes a dashboard view of the content, members, and groups in your ArcGIS Enterprise portal. This includes information such as the most popular items, a breakdown of item types, and the number of new groups added. You can also create more detailed member and item reports to gain further insight into the way people in your organization use ArcGIS Enterprise.
- **Configure settings:** The organization settings allow you to configure your ArcGIS Enterprise organization to meet your needs. Among the available settings are options to change the information displayed in item details, ArcGIS Server site federation, custom member roles, collaborations with other organizations, and security options.

As an administrator, you will likely spend most of your time using the tools on the **Organization** tab. But you will need to be familiar with the portal as a whole so you can support the other users in your organization.

Tutorial 6: Familiarize yourself with the ArcGIS Enterprise organization portal interface

In this exercise, you will navigate to the different parts of the portal. The goal of this exercise is to familiarize yourself with the interface. As part of that process, you will start asking questions about whether the portal is configured to meet your needs.

Keep track of your answers to these questions, especially if it seems that your Enterprise deployment may not currently be configured to meet all your needs. In later exercises, you will look at the workflows to make changes to the portal configuration.

Confirm your member account details

Your member account should accurately reflect your personal information, settings, and licenses.

1. In your browser, navigate to your ArcGIS Enterprise portal website.

2. Sign in to the portal.

> *Tip: Depending on how your portal is configured, you might need to click the Sign in button, or you may be automatically signed in. When your name and username are displayed in the upper-right corner of the page, you know you have successfully signed in.*

3. In the top-right corner of the page, click your profile picture. Then click **My profile**.
   - Are your profile picture, name, bio, and profile visibility correct? If not, edit any incorrect values.

4. In your **Item gallery**, review the content items.
   - Are the items that show up in your profile's item gallery the ones that you you want people to see when they access your profile? If not, click **Customize items** to change your individual gallery items.

5. At the bottom of your profile, click **View my settings**.
   - Are the email address, language, number, date, and units set correctly? If not, change any incorrect values.

6. On the left, click the **Security** tab.

   If you need to change your password or security question, you could do that from the security settings.

7. On the left, click the **Licenses** tab.
   - What is your user type and role? If your user type and role are not set correctly, refer to chapter 7 for information on how to update them.

Review the Home tab

1. On the top navigation bar, click the **Home** tab.

   > *Tip: This may or may not be the page you are automatically directed to when you first log in.*

   - Does the home page reflect the image of your organization that you want people see when they first arrive at the portal? If not, click the **Edit home page** button at the bottom to see the available customization options available in the home page editor. Consult with your organization's branding stakeholders before making permanent changes to the home page.

Review the Gallery tab

1. Navigate to the **Gallery** tab.

2. Under **Filters**, expand the **Date modified** section and click **Last 30 days**.

    This limits the gallery to layers modified in the last 30 days.

3. At the top of the gallery, click **Relevance** and change the sort method to **View count** to find the most popular items.

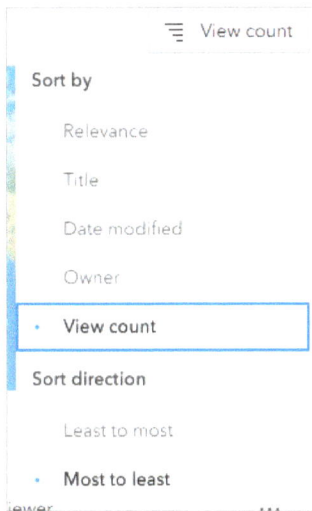

6

4. Experiment with the view and sort options and filters to show different content items in the gallery.
    *   Will users of the portal be able to use the gallery effectively to find content that is relevant to them?

Review the Map tab

1. On the navigation bar, click the **Map** tab.

    Depending on your version of ArcGIS Enterprise and how your portal is configured, the **Map** tab may bring you to either Map Viewer (first image) or Map Viewer Classic (second image).

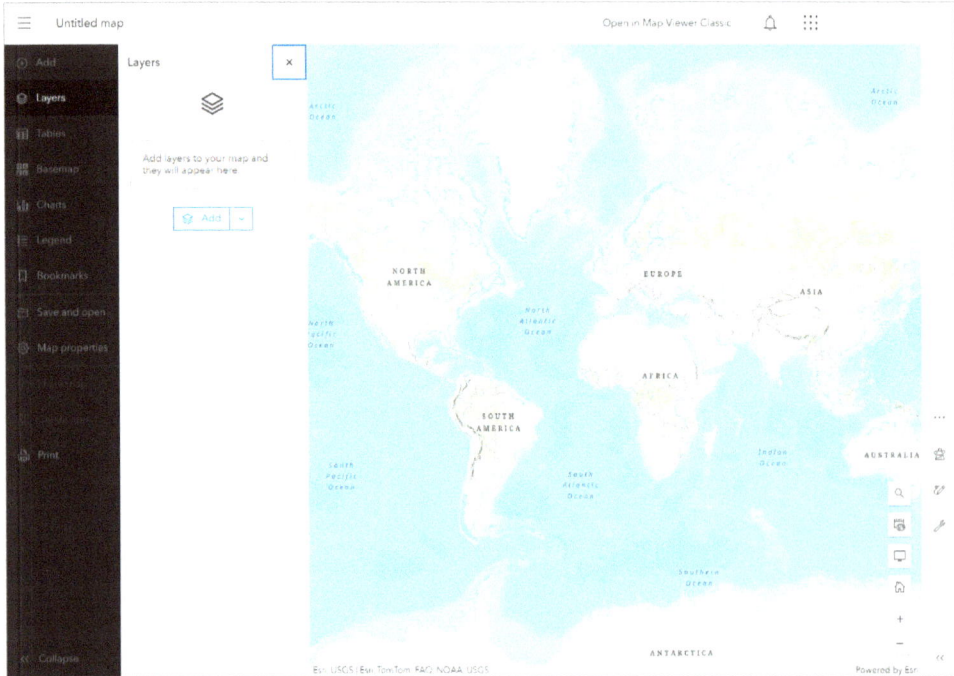

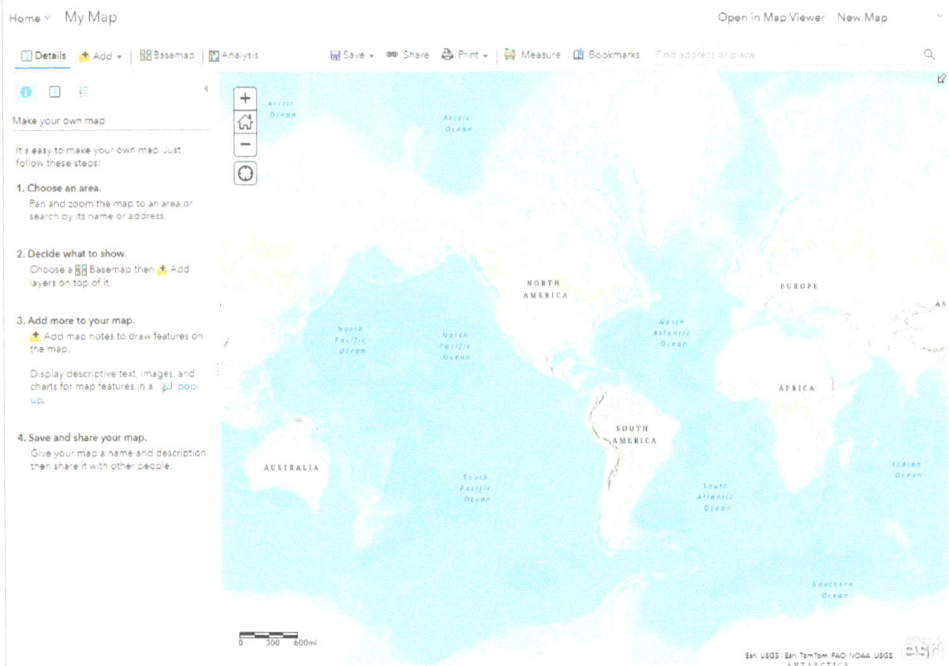

- Which app opens for you?

2. Determine whether that is the correct web mapping application.

Note: Map Viewer Classic is based on the 3.x version of the ArcGIS API for JavaScript, which was retired on July 1, 2024. Some organizations may have versions of ArcGIS Enterprise that do not support the newer Map Viewer and some may have workflows that require Map Viewer Classic to be the default web mapping application in their ArcGIS Enterprise portal. Esri recommends that all organizations transition to the new Map Viewer.

---

3. Click the **Basemap** tab. In Map Viewer, the button will be on the **Contents** toolbar. In Map Viewer Classic, it will be on the top toolbar.
   - Which basemap is used by default? Is that the correct basemap for your organization?

Review the Scene tab

1. Return to the **Home** page and click the **Scene** tab to open Scene Viewer. In Map Viewer, next to the title of the map, click the three horizontal lines and select the tab from the list. In Map Viewer Classic, click the arrow next to **Home** and select the tab from the list.

2. Choose one of the preconfigured scenes.

3. Zoom into a location and view the 3D scene.

Review the Groups tab

1. On the upper left of the screen, click the three lines and select the **Groups** tab.

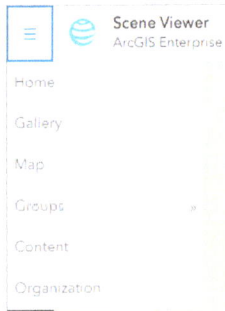

   - How many groups are you a member of?

2. If you are a member of at least one group, click **View details** for that group. If not, skip to the "Review the Content tab" section next.

3. On the group's item page, click the **Content** tab.
   - How many content items are shared to this group?

4. Click the **Members** tab.
   - How many users are members of this group?
   - Is the group membership correct? If the group membership is incorrect, see chapter 7 for information on how to add or remove members.

5. Click the **Settings** tab.
   - How can people view, join, contribute content, and see the full list of group members?
   - Are those settings correctly configured for this group?

Review the Content tab

1. On the top navigation bar, click the **Content** page.

   > Tip: This is different from the Content tab of the Groups page.

   - How many content items do you own?

   > Tip: If you want to add any frequently used items to your Favorites, click the three dots (overflow menu) for that item and select Add to favorites.

2. Click the **My favorites** tab.
   - How many items have you made favorites?

3. Click the **My groups** tab.
   - Across the groups you are a member of, how many content items are shared to those groups?

4. Click the **My organization** tab.
   - How many total items in your organization are available to you?

Review the Organization tab

1. On the navigation bar, click the **Organization** tab.

    > *Note: The Organization tab is different from the My Organization tab on the blue content details bar.*

    - How many total members are there?

2. Click the **Members** tab.

3. In the list of members, click the **Last login** header to change the sort order of members so that the members who have the least recent **Last login** are at the top.
    - Do you have any members who have never signed in?

4. Click the **Licenses** tab.

    You are currently viewing the **Add-on licenses** section for your organization.

5. Switch to the **User types** section.
    - When do your user type licenses expire?
    - How many of each user type is your organization licensed for?

6. How many of each license are assigned to a user?
    - Is that the right number of licenses?

7. Click the **Status** tab.
    - How many content items were created in the last 14 days?

8. Click the **Settings** tab. Then, in the list of tabs on the left, open the **Gallery** section.

    By default, the gallery lists all the content items in the organization.

    If you have a specific group for items that should be listed in the gallery, you can choose it here.

9. Go to the **Map** section.

    If the primary Map Viewer or basemap is not set correctly for your organization, you can change it here.

10. Go to the **ArcGIS Online** section.
  • Is your organization currently configured to use any ArcGIS Online utility services?

11. Go to the **ArcGIS Living Atlas** section.

> *Note: This section is available only in ArcGIS Enterprise versions 11.3 or later. For earlier versions, ArcGIS Living Atlas of the World content is tied to a specific version of ArcGIS Enterprise. See chapter 20 for information on upgrading ArcGIS Enterprise.*

If your ArcGIS Living Atlas content is out-of-date, you can update it here.

12. Go to the **Servers** section.
  • How many federated server sites are configured with your ArcGIS Enterprise deployment?
  • What Server roles do they have?

13. Go to the **Security** section.
  • Is your portal configured to allow access through HTTPS only?
  • Is your portal configured to allow anonymous access?
  • Are those security settings correct? If the security settings are incorrect, see chapter 4 for information on configuring security options.

Summary

In this chapter, you reviewed the ArcGIS Enterprise portal, gaining the familiarity with the portal that you will need to effectively perform the administrative tasks in subsequent chapters.

CHAPTER 7

Managing users in ArcGIS Enterprise

Objectives

- Manage licensing.
- Manage user privileges.
- Create members.
- Delete members.

Introduction

In this chapter, you'll learn how to manage the members of your ArcGIS Enterprise organization. Setting the appropriate member configurations is important because it affects who will be able to sign in to your ArcGIS Enterprise portal and what they will be able to do after they sign in.

Licensing

Each member is licensed to access certain capabilities of ArcGIS Enterprise. The primary way to assign licenses is to give a member a particular user type. Beyond their user type, members may also have add-on licenses for applications and user type extensions.

A Portal for ArcGIS license file specifies the number and variety of user types and add-on licenses you have available to assign to members. Talk with your Esri representative to make sure you have the right quantity and types of licenses for the work your members need to do.

User type

Every member of your ArcGIS Enterprise portal must be assigned only one user type. A user type defines the capabilities they have within ArcGIS Enterprise and the default set of applications that they can access and use. The specific details of what each user type enables can vary between versions of ArcGIS Enterprise but fall broadly into four categories:

- **Viewing:** Members with any user type have the capability to view content that has been shared with them by other members.
- **Editing:** Some user types provide members with the capability to edit data in content items that have been shared with them by other members.
- **Creating:** Some user types allow members to create content items. The creating capability includes making maps and apps, publishing layers, and performing analysis.
- **App access:** A member's user type will determine which apps they can access. For a given user type, these apps could include ArcGIS Pro as well as web apps, such as ArcGIS Dashboards, ArcGIS Instant Apps, and ArcGIS Experience Builder.

For example, members with the Viewer user type can view items that have been shared with them, but they do not have the capability to own, create, or edit data for those items. The Viewer user type also has the capability to use fundamental web apps, such as Dashboards or ArcGIS StoryMaps℠. Members with the Professional user type, on the other hand, have the capability to view, own, and edit content items and have access to a broader range of apps, such as ArcGIS Notebooks and ArcGIS Pro.

Check the documentation for your version of ArcGIS Enterprise so that you can assign your members the user type that best matches the capabilities and apps they require to do their work.

Add-on licenses

Some members may require access to additional apps beyond the default granted by their user type. For example, a member assigned the Contributor user type does not include access to ArcGIS Pro and would need an add-on license to use it.

When you think about the add-on licenses you require, consider the interaction between the add-on licenses and the member's user types. For example, instead of a Contributor user type and an add-on license for ArcGIS Pro, you may be able to reduce costs and administrative overhead by assigning the member a user type that includes ArcGIS Pro.

User privileges

Distinct from a member's capabilities, each member also has a set of privileges. Although privileges are not the same as capabilities, they are related, and the privileges granted to a user must be compatible with the capabilities defined by their user type. For example, the

Contributor user type enables the capability to view and edit data that has been shared with them. A member with the Contributor user type could be granted the privilege to view data. That member could not, however, be granted the privilege to own content, since that is not one of the capabilities the Contributor user type has.

Privileges are defined in two categories: general and administrative. General privileges allow members to perform common user workflows, such as creating groups, sharing content items, publishing layers, and editing data. Administrative privileges allow members to change the ArcGIS Enterprise deployment or change other members' content. These privileges include federating ArcGIS Server sites, adding members, and reassigning ownership.

When an administrator assigns privileges, an important security best practice is the principle of least privilege: A user should have access to every capability they need to do their job, but they should not have access to any capability they do not need. Members with any administrative privileges have the authority to make impactful changes to your ArcGIS Enterprise deployment. Excessive administrative privileges could cause serious problems for your organization. For that reason, an organization will typically have many more users with general privileges than administrative privileges. Although the privileges in the **General** category have less impact on the functionality of your ArcGIS Enterprise deployment, you should still adhere to the principle of least privilege when granting these privileges.

Default roles

When you first install and configure ArcGIS Enterprise, there are default roles representing bundles of privileges that are broadly appropriate for a few member personas.

- The default **Viewer** role can view content items that have been shared with them but cannot do anything else.
- The default **Data Editor** role can view and make edits to the data of content items that have been shared with them but cannot create, own, or share items.
- The default **User** role has most general privileges but cannot create or edit services.
- The default **Publisher** role has access to nearly all general privileges. Notably, the default **Publisher** role can create and edit services as well as view, create, own, edit, and share portal content items.
- The default **Administrator** role has every privilege.

There are some special considerations for the default Administrator role. At least one member must be assigned this role. It is a good idea to have more than one member assigned a default Administrator role so that administrative tasks can still be done even if one administrator is unavailable.

The specific available privileges and compatible user types for each default role can change between versions of ArcGIS Enterprise. Refer to the documentation for your version of ArcGIS Enterprise for a list of privileges and compatible user types. If a member's user type is incompatible with the role you want to assign them, you will need to change their user type to a compatible one first.

Custom roles

The default roles may not be appropriate for all members of your organization. It may be the case that a member needs to perform work that does not match any default role. For example, a member may need to create and edit Notebooks, which is a privilege that is granted by default only to the Administrator role. If that member does not require the full set of administrator privileges to do their job, you should not grant them that role. Instead, you should create a custom role that better aligns with the principle of least privilege.

Most likely, the custom role you need for a member will be similar to an existing role. You can start with that role and add or remove privileges from the full list of privileges, as necessary.

When selecting privileges for a custom role, be aware of the compatible user types for that bundle of privileges. You are not able to assign a role to a member that includes privileges that are incompatible with the capabilities included in their user type. For example, a member with the Viewer user type cannot have a custom role that includes the privilege to create, update, and delete content items, since that is not a capability of the Viewer user type. As with default roles, you may need to assign the member a new compatible user type before assigning them the custom role.

User role versus user type

People occasionally conflate *user type* and *user role*, but it is important to understand that they are different. User types are the first decision you make. It is your way to make sure you have the right user types to provide the capabilities that all your users need to do their job. Then assign each member a role with the privileges they need.

Member categories

Starting at ArcGIS Enterprise 11.1, you can organize the members of your ArcGIS Enterprise portal into hierarchical categories to help manage members by job role, department, or other characteristics. Categories are convenient for filtering and selecting members to manage in bulk. For example, if all the members of a team should be added to a particular ArcGIS Enterprise group, as an administrator, you can use categories to filter and select all those members at once instead of searching for them individually by name.

Members can belong to multiple categories. The hierarchical design of member categories means that when you add a member to a category, they are automatically added to every category above it in the hierarchy. You can set a member's categories when you create the member account, and you can set or change those categories at any time.

Creating members

To allow a user to sign in to ArcGIS Enterprise, you must add them as a member to your ArcGIS Enterprise organization. The process for adding members may depend on the authentication process you have set up to access ArcGIS Enterprise. See chapter 4 for details on authentication options.

If you are using the built-in identity store, you will specify the member's name, email address, username, user type, user role, and password. You must provide a username and password to the member, and you should require them to change that password the first time they sign in.

If you are using an external identity store, such as Security Assertion Markup Language (SAML) or Active Directory for member authentication, you can search for members in that identity store. Because these types of user accounts are separate from ArcGIS Enterprise, the name, email, username, and password are already configured. However, you must still configure a user type and compatible user role.

External identity stores may also have predefined groups of users, which you may be able to use to help add members. For example, if you use Active Directory as the external identity store, you can add members in bulk by their group. Members who are added this way will have the same user type and user role when they are first added.

Regardless of the authentication scheme you have configured for ArcGIS Enterprise, you can add members in bulk by specifying the relevant information in a text file. When you add members this way, the ArcGIS Enterprise portal will provide a CSV template you can download with the required fields for bulk member creation.

You can also configure ArcGIS Enterprise to allow users to add themselves as members if you configure new member defaults in the ArcGIS Enterprise portal organization settings. Each member will create their account the first time they sign in to the organization site. Be careful when using this option because you may rapidly run out of available user type licenses. Generally, it is best to use the Viewer user type and Viewer user role as defaults and update the user type and user role for members who require additional capabilities.

One good use case for having members add themselves is when you are not sure who in your organization will want access to ArcGIS Enterprise. Allowing members to add themselves when needed means that you don't have to consume a license until a user actively starts using the capabilities of ArcGIS Enterprise.

Fictional user story

Every summer, Medio County hires about 15 interns who work across county departments. Everybody who works for the county requires some level of access to capabilities in the county's ArcGIS Enterprise deployment, and interns are no different. Elise Medina, the county GIS manager, is responsible for making sure the interns can access appropriate capabilities, apps, and privileges necessary to do their jobs.

Some interns will work directly for Elise in the GIS department. These interns will perform analysis, manage data, and publish data to ArcGIS Enterprise, mostly using ArcGIS Pro. Elise assigns these interns the Professional user type. She has learned from experience over the years not to grant interns the full privileges of the default Publisher role, so she assigns them a custom Intern Publisher role that disallows unnecessary privileges such as publishing in bulk and sharing content with the public.

For Parks Department interns who will collect and update data using ArcGIS Field Maps, Elise assigns them the Mobile Worker user type and the Data editor role. For the other interns, Elise assigns them the Viewer user type and Viewer role.

As part of the member creation process, Elise assigns all the interns to the Intern member category.

Deleting members

You may occasionally need to delete members. This might happen if a user leaves your organization or if their job role changes in a way that means they should no longer have access to content or ArcGIS Enterprise organization resources. When you delete a member, their licenses are revoked and made available for you to reassign. You can delete members individually or in bulk.

Members who own content items or groups cannot be deleted while they still own those items or groups. For this reason, members who were added automatically from an external identity store such as Active Directory are not automatically removed if they are deleted from that identity store.

It may take some time to determine what should happen to the groups and content owned by members you are deleting. While you make that determination, you can quickly disable a member's account to prevent them from signing in or using an organization's resources. Disabling accounts is also a good option for members who should have their access to capabilities temporarily removed, such as people who are on leave.

Fictional user story

At the end of the summer, Elise needs to delete the interns' member accounts from the ArcGIS Enterprise portal.

Three of the interns from ArcGIS Enterprise have been hired into full-time positions with the county. From the Members tab of the Organization page, Elise selects these three members and recategorizes them to remove the Intern category.

Each departing intern was supposed to delete any unneeded content or groups and transfer the remainder to a new owner, but Elise knows from experience that sometimes that doesn't happen. Before deleting any members, she runs a Python script that transfers any groups owned by any user in the Intern member category to herself.

After running the script, Elise selects the remaining members of the Intern category and deletes them in bulk, choosing the option to transfer their content to herself. After the interns' accounts are deleted, she will go through the leftover content and groups she now owns and reassign any that should have a different owner.

7

Tutorial 7: Create a new member with appropriate privileges

In this tutorial, you will practice the workflows for managing members. You will create a member, investigate available licenses and roles, create and assign a custom role, create and assign a member category, and delete the member.

This tutorial involves making changes in your ArcGIS Enterprise portal. The changes in this tutorial are low risk because they do not involve your current members, content, or ArcGIS Enterprise configuration. It is still recommended, however, that you perform the steps of this tutorial in a test or development environment to avoid any risk of inadvertently making undesired changes to your production environment.

Create a new member

1. Sign in to ArcGIS Enterprise with an account assigned the default **Administrator** role or a custom role with appropriate privileges.

2. Navigate to the **Organization** tab.

3. Click the **Members** tab and then click **Add Members**.

4. Choose the option to **Add built-in portal members** and then click **Next**.

5. Click **New member**.

6. Complete the **New member** information with the following values:
   - **First name**: Test
   - **Last name**: Viewer
   - **Email address**: test.viewer@invalid.internal
   - **Username**: test.viewer
   - **User type**: Viewer
   - **Role**: Viewer
   - **Password**: test.viewer1

   Note: You must have at least one available Viewer user type, and the assigned username cannot belong to any existing member.

Member	User type	Role	Included licenses	
Test Viewer test.viewer test.viewer@invalid.internal	Viewer	Viewer		✏

7. Click **Next** and then continue to accept the default values for other member properties.

8. Click **Add members**.

View options for changing member licensing and user role

1. From the **Members** tab, click the three horizontal dots (menu button) for the Test Viewer member you just created and then click **Manage user type**.

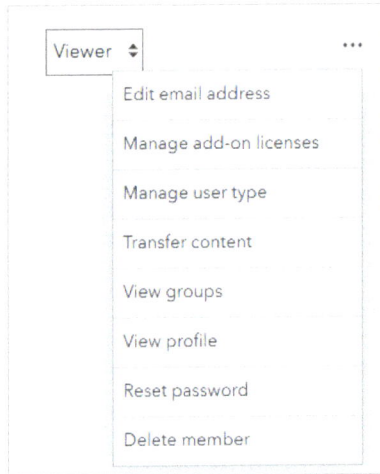

2. Under **Role**, click the **Viewer** role to see that there are limited compatible roles available with the **Viewer** user type you assigned this member.

3. Under **User type**, click the **Viewer** user type to open the drop-down menu. From the list, select **Creator**.

4. Open the **Role** drop-down menu again to see the expanded list of available roles that are compatible with the **Creator** user type.

5. Click **Cancel** to close the window without saving any changes.

6. Click the menu button for the **Test Viewer** member and click **Manage add-on licenses**.

7. Review the list of available add-on licenses, if any.

   Tip: Because this user has the lowest level of user type, most add-on licenses will be unavailable to assign.

8. Click **Cancel** to close the window without saving any changes.

Manage a custom role

1. Click the **Settings** tab. On the left sidebar, open the **Member roles** section.

2. Click **Create role**.

3. Set the following parameters for the new role:
    *   **Role name**: Limited Viewer
    *   **Description**: Role with reduced viewing privileges
    *   **Privilege compatibility**: View

4. For **Role privileges**, click the option to **Set from existing role**.

5. For **Select role or template to import settings from**, select **Viewer** and then click **Import settings**.

6. Click the **Expand all** button to see all role privileges.

7. Under **Groups**, turn off the privilege to **Join organizational groups** and **View groups shared with organization**.

8. Under **Content and Analysis**, turn off the privilege to perform **Geocoding** and **Network Analysis**.

9. Click **Save**.

10. On the **Members** tab, for the **Test Viewer** member you created, click the **Role** drop-down menu and select **Limited Viewer**.

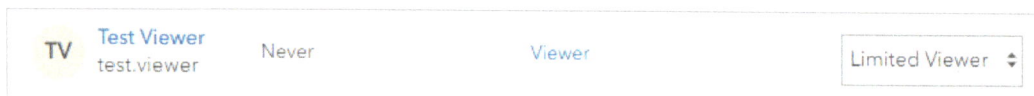

TV	Test Viewer	Never	Viewer	Limited Viewer ⬍
	test.viewer			

> *Tip: This custom role is available because it is compatible with the capabilities licensed for the Viewer user type.*

11. On the **Settings** tab, navigate back to the **Member roles** section.

12. Next to the **Limited Viewer** role you created, click the menu button and click **Delete**.

You should see an error message indicating that you cannot delete a role that has been assigned to a member.

13. Using what you have learned, reassign the **Test Viewer** member the default **Viewer** role and delete the custom **Limited Viewer** role.

Manage a member category

1. Navigate to the **Members** tab. Under **Filters**, in the **Categories** section, click **Set up member categories**.

2. In the **Category** name text box, type **Test Category** and click the **Save** button (check mark) to save the category.

> *Tip: If your organization already has member categories configured, complete the first two steps by clicking the pencil icon to edit these categories and then click the plus button to create a new category.*

3. Close the **Configure member categories** window.

4. Next to the **Test Viewer** member, check the box to select it.

5. Above the list of members, click the **Categorize** button.

6. Check the box for **Test Category** and then click **Done** to assign this category to the member.

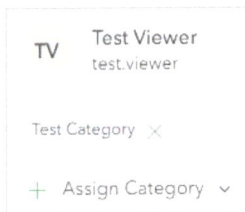

Update a member's profile and settings

1. From the **Members** tab, click the three horizontal dots (menu button) for the **Test Viewer** member and then click **View profile**.

2. Click the default profile image above the **Test Viewer** name.

3. Upload any image file to change the profile image.

4. Adjust the pan and zoom of the image and then click **Save**.

5. Click the pencil icon next to the **Test Viewer** name and change the name to any new name.

6. In the **Profile visibility** drop-down menu, change the visibility to **Private**.

7. Click **View settings** link.

8. Under **Email address**, click the pencil icon and change the email address to any new email.

9. Under **Start page**, change the value to **Content**.

10. If supported by your version of ArcGIS Enterprise, ensure that the **Primary map viewer** is set to **Map Viewer**.

Create a member report

1. Return to the **Organization** tab. From the **Status** tab, go to **Reports**.

2. Click **Create report** and then select **Single report**.

3. For **Report type**, select **Member**.

4. For **Name**, accept the default report name.

5. Click **Create report** and wait for the report to appear in the list.

6. Click the title of the newly created report to open its item page and then click **Download**.

7. Open the downloaded file and review the report.

Delete a member

1. Return to the **Organization** tab. From the **Members** tab, click the three horizontal dots (menu button) for the **Test Viewer** member and then click **Delete member**.

2. Click **Delete Member** again to confirm.

    Because this user does not own any content or groups, there is no option to transfer or delete content and groups.

7

Take the next step

Review the members, licenses, roles, and member categories in ArcGIS Enterprise. Ensure that the following values are appropriate:

- Members
- User type assignment
- Add-on license assignments
- User role assignment
- Custom role privilege configuration
- Member category configuration
- Member category assignments

The member report can be a useful tool for reviewing all your members in a single file. If you have an organization with too many members to manually review, you may want to consider automating the review using the ArcGIS Portal Directory REST API or the ArcGIS API for Python.

Summary

In this chapter, you learned the important considerations for key workflows to manage members of your ArcGIS Enterprise portal. In the tutorial, you created a member, configured their licenses and privileges, and deleted a member.

CHAPTER 8
Managing content in the ArcGIS Enterprise portal

Objectives

- Manage content sharing levels.
- Manage groups.
- Make content items discoverable.
- Manage content ownership.
- Delete content.
- Identify content item types.

Introduction

In this chapter, you'll learn about the content management capabilities of ArcGIS Enterprise. As an administrator, you do not have to have sole responsibility for content management. That work can be distributed throughout your organization, performed by the people who are most familiar with the data and purpose of each content item.

But administrators have content management privileges that most members do not. Throughout this chapter, we will refer to default administrators as those people with all the administrator privileges. You can also create custom roles with some subsets of those privileges that allow other members to perform many of the administrative tasks described in this chapter. See chapter 7 for more information on creating custom roles.

If you already know how to manage content in ArcGIS Online, the concepts of this chapter will be familiar. In places where they are different, we've highlighted those differences so you can note them easily.

Content types

A content item is an individual resource. Most content items fall into one of six categories:

- **Apps** are applications that users interact with to do their work. Most commonly, these are web apps created with app builders such as ArcGIS Dashboards, but they can be custom applications as well.
- **Maps and scenes** are collections of geographic information that are thematically connected. Users generally access maps and scenes through an app, such as Map Viewer or ArcGIS Instant Apps.
- **Layers** are representations of geographic data, such as a feature layer for vector data or an imagery layer for raster data. Maps and scenes are generally made up of one or more layers.
- **Data stores** enable access to the geographic data represented in layers. These data stores might be enterprise geodatabases, folders, or cloud stores.
- **Tools** enable analytic functions, such as creating a buffer or calculating drive times. Tools typically take layers as inputs and create layers as outputs.
- **Files** are uploaded and stored in the portal content directory. Generally, you will do that as part of the process for creating layers, such as uploading a CSV file to publish as a feature layer. You can also upload files that you would like to make available for download.

For the most part, the concepts and workflows for managing content are agnostic to the type of content item. You use the same strategies for managing a map item that you use for managing a layer item, a tool, or any other kind of item.

Item types matter most when items are related to each other. If you share a map item without also sharing all the layer items referenced by the map, people will not be able to see the information the map is intended to convey. If you delete a layer that is frequently used as an input to a tool, that tool becomes less useful. ArcGIS Enterprise will often prompt you when it detects that you are making a content management decision that implicates related content items, but it isn't able to discern all such relationships. You should be aware of those relationships so you can make good choices about content management.

Item details

Each content item has an item page that summarizes the item. It is important to add enough information to the item page so that users can understand how to use the item effectively. Item owners and default administrators will see an item information section that measures the level of detail and provides suggestions on additional detail to add. You can add and edit information for the item, such as the title, thumbnail, summary, terms of use, and several other details.

- In addition to this user-configurable information, an item page also includes automatically generated information about the item's size, as well as creation and update dates to help users better understand the item.
- You can also configure additional metadata for tracking information, such as the contact information, update frequency, theme, and more. You can enable these additional metadata fields from the organization's item settings by choosing a metadata standard, which provides access to a metadata editor from the item page.

8

Content ownership

The centerpiece of your content management strategy should be to ensure that items have the proper access permissions. Just as with any other capability, the principle of least privilege applies. You want to ensure that everybody has access to the content they need, and nobody has access to content they don't need.

The first consideration for managing access permissions is content ownership. Every content item has exactly one owner. The owner has full privileges on that content item. They can change it, share it with others, and delete it. Generally, the item's owner will be the person who is responsible for managing that item. Because the owner has significant control over the item, you want to make sure that the right person owns every item.

Some members of your ArcGIS Enterprise organization may not be able to own content. Certain user types do not have the capability to own content, and some user roles may not have the correct privileges to own content. You may need to change a member's user type or role to enable them to own content. See chapter 7 for more information about user types and roles.

Regardless of ownership, default administrators also have full privileges on every content item, including the additional privilege of changing any item's owner. You can grant other members the privilege to change the ownership of their own items, but only the Administrator role has this privilege by default.

Sharing levels

One way to set the accessibility of content items for other ArcGIS Enterprise users is by configuring the item's sharing level. There are three possible sharing levels: Owner, Organization, and Everyone.

- **Owner**: Access is granted only to the item's owner and default administrators. This is the default.
- **Organization**: Every member can access the item.
- **Everyone**: Every user who can access ArcGIS Enterprise can access the item, even if they are not a member.

If an item's sharing level makes it accessible to users other than the owner, those users can

- search for the item,
- access the item page,
- share the item to a group where the user has content sharing privileges.

If the item supports it, a user with access can also

- edit the data for an item,
- use the item in a map or app,
- download the item.

An item owner or default administrator can always change the sharing level, but if an item were inappropriately shared with too many people, information may have moved outside the control of ArcGIS Enterprise before the sharing level could be set correctly. For example, a user who should not have permission to access the item could download it and move it to another system. For that reason, it is safer to err on the side of sharing more restrictively, especially for sensitive information.

Groups

Another way to make content accessible to other members is to share it with a group. Members of the group will be able to access the item, even if the item's sharing level is set to **Owner**.

Group membership

Just like content items, every group has exactly one owner. Owners have control over the group settings, the content that is shared to the group, who is allowed to be a member of the group, and group deletion. The initial owner of a group is the member that created it. Only ArcGIS Enterprise organization members with the proper administrative privileges can reassign ownership of existing groups and only to new owners with the correct privileges for owning that group.

Groups can also have managers who help the owner with group management tasks. Group managers have many of the same capabilities to manage the group as the owner. Group manager privileges include the ability to delete the group and add or remove other group managers.

Members of the group who are not the owner or manager can always see the group and the content items shared to it. Depending on the settings configured by the group owner or managers, these other members may be able to share content items with the group and see the list of other group members.

Note: Unlike ArcGIS Online, you cannot add members from other organizations to an ArcGIS Enterprise group. All members of the group must be members of the same ArcGIS Enterprise organization. See chapter 15 for information on how you can share content between organizations with ArcGIS Enterprise.

8

When you add members to a group and set their group role (owner, manager, or member), make sure that you are giving the right level of access to the right users. Because the appropriate privileges for group membership may change over time, you should also periodically review those privileges. For organizations with many groups, you might delegate that responsibility to group owners and managers.

Group settings

A group's settings control how members can view, join, and access content shared to the group. You configure the group's settings when you first create the group, and most of those settings can be changed afterward if necessary.

When the existence of the group does not need to be widely known, you can restrict who can find the group through search and see the contents of the group page. You have three options for configuring who can view the group:

- **Only group members** means that users must already be a member of the group to be able to see it.
- **All organization members** means that anyone who can sign in to the ArcGIS Enterprise organization can see the group.
- **Everyone** means that all users who can access your ArcGIS Enterprise organization can see the group. This is the default.

Initially, the group owner is the only member of the group. You have four options for configuring how people can join the group:

- **By invitation** means that the group owner or manager must initiate the process by sending an invitation. This is the default and is a good choice when you know exactly who should be a member of the group. Default administrators can add members by invitation without requiring confirmation from the member.
- **By request** means that anybody who can view the group can ask to be added as a member by the group owner or managers. This is a good choice when you are not sure who should be a member but need to be careful about who has access.
- **Being a member of an Active Directory group** is available if your ArcGIS Enterprise portal is set to use Active Directory for the group store configuration. Members will be added and removed from the ArcGIS Enterprise group automatically based on their membership in the Active Directory group.
- **By adding themselves** means that anybody who can see the group can add themselves without approval from the group owner or managers. This is a good choice when you are not sure who should be a member, and anybody who wants access to group content can do so.

Groups can also limit who contributes content. By default, all group members have the capability to allow the group to access shared content, but you can configure the settings to allow only the group owner and managers to add content to the group.

Group designations

When creating a group, default administrators also have the option to create special group designations. A shared-update group grants all the members of the group the privilege to edit items shared to the group. This is a good idea if the group members need to collaborate on creating or updating items. An administrative group prevents a member from leaving the group unless the group owner or manager removes them. Administrative groups are used primarily to organize members for bulk member management rather than to share content.

Unlike other group settings, special designations cannot be changed after group creation. If you realize later that a group should have been a shared-update group, you will need to create a new group with that designation.

Relationships between setting options

Group settings have logical interrelationships that must be satisfied for ArcGIS Enterprise to consider the settings valid. If only group members can view a group, the only option to join the group is by invitation. Because shared-update groups grant significant editing privileges on all items shared to the group, users cannot join this type of group by simply adding themselves. Because the purpose of administrative groups is to facilitate member management, users cannot add themselves to those groups either.

In addition to invalid combinations of group settings, you should carefully consider certain combinations of valid but problematic settings and content. If people can join the group by adding themselves and all group members can contribute content, there is little administrative control over what items are shared to the group. If any items shared to the group have editable data and people can join the group by adding themselves, there is a high risk of incorrect data edits by members who should not have been granted access to that item.

8

Group sharing versus sharing level

Group sharing is parallel to sharing level:

Owner	Organization	Everyone
Group		

Sharing an item with a group is independent of the item's sharing level. The access users have for an item is the result of the combination of group sharing and sharing level (table 8.1).

Table 8.1. Group sharing and sharing level

Sharing level	Item not shared with any groups	Shared with at least one group
Owner	Only owner and default administrators can access the item.	Members of the group have most access privileges for the item, but they cannot share the item with other groups.
Organization	Every member can access the item, including sharing the item to groups where they have the privilege to do so.	No additional access.
Everyone	Every user can access the item, including sharing the item to groups where they have the privilege to do so.	No additional access.

Fictional user story

At SuperBiz, vehicle tracking information is managed by different teams depending on the type of vehicle. This division of labor has enabled each team to develop expertise in a single transportation modality. This division has also had the unfortunate side effect of making it more difficult to understand whether different modalities are integrating efficiently. Harjeet Singh, the SuperBiz director of data, has initiated a project to develop a single application that will integrate information across modalities.

Because SuperBiz has been using ArcGIS Enterprise to manage all its spatial data, the relevant datasets already exist as portal content items. Harjeet has the datasets inventoried and notices that the sharing is inconsistent and does not enable effective integration. Some items have been shared across the organization, some have been shared with groups for each vehicle team, and some are currently not shared at all. Alarmingly, some highly sensitive information has been shared more broadly than is appropriate.

Because Harjeet has the default Administrator role in ArcGIS Enterprise, he could change the group sharing of items without the owners' permission. But it is better for the item owners to change the sharing level themselves, so they are aware of all the settings on their items. After the owners have limited the sharing of sensitive items, Harjeet next creates a new group for the integration project. Because the members of the group won't be changing any existing content items or using the group for member management, it does not need to have a special group designation as a shared-update or administrative group.

Currently, the content item owners are a large group of people, and Harjeet does not think all of them should have visibility into all the items that need to be shared to this new group. Neither does Harjeet want the day-to-day responsibility of owning each of these items himself. Instead, Harjeet coordinates with the leads for each vehicle team so that all relevant items for a single team are owned by that lead. After the ownership has been changed, Harjeet invites each team lead to the group, and they share the appropriate items.

Finally, Harjeet invites two application developers to the group. They will have access to the data items while they develop the integration application. When they are finished, Harjeet removes them from the group because they no longer need that access.

Discoverability

It isn't enough that users have privileges to access content. They must be able to find that content. Content discoverability can be especially challenging when an organization has hundreds or thousands of content items that users would need to scroll through to find what they are looking for.

One strategy is to use groups to help people discover content. Even if the item sharing level is already set to Organization or Everyone, items can be easier to find when a user needs to look through only a small number of items in a group rather than a large number of items in an entire organization.

But using groups to aid discoverability only helps if you already know the members and content that should be added to certain groups. For most use cases, you will want to make sure the item itself has enough information that users can use the searching and filtering functions of the ArcGIS Enterprise portal to locate the content items they need.

Content organization

Groups aren't the only way to organize content. You can also add tags to an item's details so that the item turns up in a search that uses any of those tags. Because information from the item's title, summary, and description are also used in a search, you do not need to duplicate that information in a tag. Tags are free text, so a user with the ability to modify an item can add anything they want as a tag. But the best way to use tags effectively is to have a relatively small number of specific tags.

Instead of free-form tags, you can also use more structured content categories to organize items. From the organization's Item settings, you can configure a hierarchy of categories and subcategories. Use categories instead of tags when you need to make sure that items are organized in a specific predetermined pattern. Content categories can help filter content items and aid discoverability.

Searching

Using filters such as Categories can narrow the number of items, but users will still need the ability to search for the specific items they need. On the Content tab of the organization site, users can type in the search bar to find matching content items. For example, the search term *outages* will return all content items that match the word "outages" in their Title, Summary, Description, and Tags.

Sophisticated users can also specify Boolean operators and designate exact search fields to find matching content. For example:

- The search term *outages AND electric* will return all content items that match both "outages" and "electric."
- The search term *outages NOT gas* will return all content items that match "outages" but do not match "gas."
- The search term *owner:portaladmin outages NOT gas* will return all content items owned by the member with the username portaladmin that match "outages" but do not match "gas."

The most important thing you can do to facilitate users' ability to search effectively is to provide complete information on the item page. Blank or incomplete item details mean search terms won't return all relevant content items.

In addition to searching by item details, if an item has a spatial extent, you can also enable searching by that extent. Spatial search is important if a user wants to find items that are relevant to a specific geographic location. If your organization has many content items that have similar information for different areas, enabling search by extent can help users find the right information for their area of interest.

Content status

A well-used ArcGIS Enterprise organization will contain many content items, but not all those content items will reflect best practices or the most up-to-date data. From an item's settings, you can set the content status to help users know if they should use a particular item or not. Marking an item as authoritative means that you recommend the use of that item. Marking an item as deprecated means that you discourage the use of that item. When looking for items, users can filter by status, and both authoritative and deprecated items will have a status badge icon that displays next to the content item. Content status helps aid item discoverability by steering users toward authoritative items and away from deprecated items.

Content apps

Complex search techniques are not the right choice for all users. If your goal is to make it as easy as possible for users across your organization to find the content they need, you will need more than just the ability to search. You can make content items more easily discoverable by creating curated content applications.

One way to create a curated collection is to use the gallery that is built into your ArcGIS Enterprise organization site. By default, the gallery shows a list of all organization items, but you can specify a group instead. When you share selected items with that group, you narrow the list of content items users see when they navigate to the gallery, which enhances the discoverability of those items. Items that are widely accessed by a diverse range of users and items you want to promote are good candidates to add to the gallery.

ArcGIS Enterprise contains a single gallery, so you generally want to use it to highlight content that is relevant to users across your organization. To enhance discoverability for content with a narrower scope of interest, you can also create separate gallery apps that have focused collections of content. The easiest way to create such an app is to use the Category Gallery Instant Apps builder. From any group, you can create an app that highlights the content shared to that group. This is especially good for users who do not need to access the full functionality of the ArcGIS Enterprise portal but just want a single app to access relevant content.

If you need more capabilities than a simple collection of content items, you can use ArcGIS Enterprise Sites to create a website to feature content. Sites include links to content and integrate dashboard elements, such as maps and summaries, to bring the relevant content information to a focused experience. Sites are particularly useful for making content usable for people without a GIS background but who still need access to information made available through ArcGIS Enterprise.

> *Tip: ArcGIS Enterprise Sites are similar to ArcGIS Hub Sites used with ArcGIS Online, but they feature the content items in your ArcGIS Enterprise organization.*

Content folders

At the other end of the spectrum, some power users of ArcGIS Enterprise will own so many content items that it is difficult for them to keep track of their own items. In addition to ensuring quality item details to facilitate search, users can also categorize their own content items by putting them into folders. Folders in the ArcGIS Enterprise portal don't affect how items are stored on disk, nor do they affect how other users interact with the items. Folders are virtual directories that exist only to organize content items. Users may find it convenient to organize their items by project, data type, or any other categorization schema that makes sense to them.

8

Fictional user story

During the development of SuperBiz's multimodal transportation integration app, analysts testing the app discovered that it is missing some important information. Harjeet, the director of data, reviews the content item inventory that was performed at the beginning of the project. The review reveals that several content items did not turn up in the search because their item details were incomplete or incorrect.

Before the data integration project, the people who owned and used the content items had intimate knowledge of which items were used for which purposes. Now, however, these items need to be shared with other people without that high degree of familiarity. Short or missing summaries and descriptions, lack of tags, and cryptic item names made it impossible for other analysts at SuperBiz to effectively find the information.

Investigating the issue further, Harjeet discovers that this problem is widespread, not limited to the content items needed for the integration project. From the content owners, Harjeet learns that they are under enormous time pressure to produce results quickly, which makes it challenging to ensure that content items also have sufficient item details. They simply don't have enough time to do both.

Because fast results and detailed item information are crucial business needs, Harjeet decides to add a new Data Librarian role to his group. Content owners have an initial responsibility to complete item information but won't be held accountable for ensuring complete accuracy. The Data Librarian will review items, correct any inaccuracies, and complete the item details according to SuperBiz's standards.

Deleting content

One of the main barriers to content discoverability is having too many items, which makes it difficult to distinguish between relevant and irrelevant content. Too many items may also put pressure on your storage resources. At the same time, some of your items represent content of irreplaceable value, so you cannot be too aggressive about deleting them. Consequently, it is vital to have good strategies for managing content deletion.

An item can be deleted by its owner or by a member with the administrative privilege to delete content. Generally, individual content owners will be responsible for deleting their content items, but administrators may need to periodically remind owners to delete any unnecessary items.

Your deletion strategy is affected by relationships between items. Sometimes, an item will be dependent on another item, which prevents deletion. For example, a hosted feature layer view is always dependent on an underlying hosted feature layer. ArcGIS Enterprise will not allow deletion of the hosted feature layer until the view is deleted. Sometimes items will be related but not have a dependency, which does not prevent deletion. For example, if a map item contains a layer item, ArcGIS Enterprise will neither stop nor warn you if you delete the layer item. But the map will be broken because it relies on a layer item that no longer exists.

To prevent accidental deletion of important content items, the item owner or a default administrator can turn on delete protection. If delete protection is turned on, an item cannot be deleted. Delete protection won't prevent purposeful deletion, because it can be turned off at any time by the owner or default administrator, but enabling it is a good idea for important content.

Note: Unlike ArcGIS Online, ArcGIS Enterprise does not support a recycling bin for easily restoring deleted items. See chapter 18 for information on using backup and restore processes to gain access to deleted items.

Tutorial 8: Use ArcGIS Enterprise to manage content items

In this tutorial, you will practice the workflows for managing content items. You will create items, configure their settings, investigate relationships between items, change ownership, create a group, change the sharing of items, and delete items.

This tutorial involves making changes in your ArcGIS Enterprise portal. The changes in this tutorial are low risk because they do not involve your existing members, content, or ArcGIS Enterprise configuration. It is still recommended, however, that you perform the steps of this tutorial in a test or development environment to avoid any risk of inadvertently making undesired changes to your production environment.

Create content items

First, you will create some new content items to manage.

1. Create a .csv file and populate the file with the following text:

```
LAT,LON
0,0
```

2. Save the file as **null_island.csv**.

3. Sign into your ArcGIS Enterprise portal. Navigate to the **Content** tab and click **New item**.

4. Drag the **null_island.csv** file to the window to upload it.

5. Select the option to **Add null_island.csv and create a hosted feature layer or table** and then click **Next**.

6. Accept the default options for the field types and then click **Next**.

7. For **Location settings**, ensure the field for **Latitude** is **LAT** and the field for **Longitude** is **LON**. Click **Next**.

8. Accept the remaining defaults and click **Save**.

9. Navigate to the **Content** tab.

   Notice how you have two items made from the same file—a CSV file and a hosted feature layer.

☐ null_island	🔲	Feature Layer (hosted)
☐ null_island	🔲	CSV

Configure content items

These content items need to be configured to fully document information about them and to enable the correct capabilities.

1. Click the **null_island** hosted feature layer to open its item page.

   You will see that the green bar on the right under **Item Information** indicates there is not enough information about this item and the recommendation to add a summary.

2. Next to the **Summary** section, click the **Edit** button. Type the following summary:

   Point layer documenting the location of Null Island in the Atlantic Ocean.

3. Next to the **Description** section, click the **Edit** button and add the following description:

   When the location of spatial data is unknown, these data are often geolocated to an arbitrary location. Most commonly, this arbitrary location is at 0 degrees latitude and 0 degrees longitude. Because this happens so frequently, people who work with spatial data have termed this location "Null Island."

   Data can also end up on Null Island when they are interpreted in the wrong spatial reference. For example, a point that should be located at 45 degrees North latitude and 90 degrees East longitude might end up 45 meters north of 0 degrees latitude and 90 meters east of 0 degrees longitude.

4. On the right of the page, scroll down to the **Tags** section, click the **Edit** button, and add the following tags:

   Tip: You can create multiple tags by separating them with a comma.

   - Getting to Know ArcGIS Enterprise
   - Tutorial
   - CSV

5. Click **Save**.

6. Next to the **Terms of Use** section, click the **Edit** button and add the following disclaimer:

   Fictional data. For educational or training purposes only.

7. At the top of the item page, click the **Settings** tab to open the item settings.

8. Under **Content Status**, click **Mark as Deprecated** to discourage the use of this item.

9. Under **Delete Protection**, check the box for **Prevent this item from being accidentally deleted**. Save the updated settings for the **General** section.

10. Scroll down to the **Data Source** section. Under **Editing**, check the box for **Enable editing**. For **What kind of editing is allowed?**, uncheck the box for **Delete**.

11. Save the updated settings for the **Data Source** section.

8

Create related items

To see how items can be dependent on each other, you will create a hosted feature layer view and a web map.

1. Click the **Overview** tab to return to the item page for the **null_island** hosted feature layer.

2. On the right side of the page, click **Create View Layer** and select **View Layer**.

3. Click **Next** to accept all defaults and then click **Create**.

4. Navigate to the **Settings** tab for the newly created hosted feature layer view.

5. Mark the item as deprecated and ensure the **Enable editing** capability is turned off.

6. On the navigation bar, click the **Map** tab to open Map Viewer.

7. In the **Layers** pane, click **Add**. From **My content**, add the **null_island view** item to the map.

8. On the **Contents** toolbar, click **Save and open** > **Save as**. Name the map **Null Island** and apply the same **Tags** and **Summary** as the **null_island** hosted feature layer. Click **Save**.

9. Return to the **Home** page and click the **Content** tab.

10. Click the **Null Island** web map to open its item page and verify that the **null_island view** item is listed as a layer in the web map.

Change item ownership

1. Create a new member with the appropriate user type and role to own content items, such as a **Creator** user type and **Publisher** user role.

   > *Tip: See chapter 7 for information on how to create new members. If you are comfortable affecting other members as part of this tutorial, you can also use an existing member who can own content items.*

2. Navigate to the item page for the **Null Island** web map item you created.

3. On the right side, click the **Change owner** button and select the new member you created (or existing member if you did not create a new member).

4. Navigate back to your list of content and confirm that you no longer own the web map item, but you are still the owner of the CSV, hosted feature layer, and hosted feature layer view items.

Create groups

1. On the navigation bar, click the **Groups** tab. Click **Create group** and apply the following settings for a new group:
   - **Name**: Null Island Editors.
   - **Summary**: To control access to editing data related to Null Island.
   - **Tags**: Getting to Know ArcGIS Enterprise, Tutorial.
   - **Who can view this group**: Only group members.

2. Accept all other defaults and click **Save**.

3. Create another group with the following settings:
   - **Name**: Null Island Viewers.
   - **Summary**: To enable members to easily locate data related to Null Island.
   - **Tags**: Getting to Know ArcGIS Enterprise, Tutorial.
   - **How can people join this group**: By adding themselves.
   - **Who can contribute content**: Group owner and managers.

4. Accept all other defaults and click **Save**.

5. On the Null Island Viewers group **Overview** tab, click **Invite members**.

   At the bottom of the window, notice the check box that allows you to add organization members without requiring confirmation.

6. Select the member who is the owner of the Null Island web map and click **Add members to group**.

7. Click the **Members** tab of the group.

8. Select the new member and then click **Update member's group role**.

9. Change their role to **Manager** and then click **Save**.

Configure sharing

1. On the navigation bar, click the **Content** tab. Open the item details for the **null_island** hosted feature layer.

2. Click **Share** and then click **Edit group sharing**.

3. Share this item with the **Null Island Editors** group and then click **Save**.

4. Navigate back to the **Content** tab and then go to **My Organization**.

5. Search for the **Null Island** web map and then click its title to open the item page.

6. Click **Share** and then click **Edit group sharing**.

   Tip: In this exercise, you will change the sharing settings for another user's item to see that it is possible, but it is best practice for item owners to change the sharing on their own items.

7. Select the **Null Island Viewers** group and then click **Apply**.

8. For **Set sharing level**, select **Organization** and then click **Save** to confirm the sharing changes.

   Notice that ArcGIS Enterprise automatically detects that the web map item is being shared differently than the view layer it contains.

9. Click **Review sharing** and then click **Update sharing** to match the sharing settings of the view layer to those of the web map.

   Because the web map and view layer are shared with the entire organization, adding the items to the group did not give group members any access to those items. But because the group contains only two items, they will be easier to find than having to search the entire organization's content.

10. Navigate to the **Groups** tab. Find the **Null Island Editors** group and click **View details**.

11. On the right, click **Add items to group**.

This is an alternative method of adding items from the Groups page, instead of from the Content page.

12. Select the **null_island** view layer and then click **Add items**.

The effect of creating a noneditable hosted feature layer view shared with the organization and sharing the underlying editable hosted feature layer with a group only is that you can limit who can edit the data without setting the same limits on who can view the data. That is a useful pattern for ensuring adherence to the principle of least privilege.

Delete items

1. From the **Groups** tab, delete the **Null Island Editors** and **Null Island Viewers** groups.

2. Navigate to the **Content** tab.

3. Search for items related to **Null Island** and confirm that none of the content items were deleted just because the group they were shared with was deleted.

4. Select the **null_island** hosted feature layer and then click **Delete**.

This item cannot be deleted because it has delete protection and has a dependent hosted feature layer view.

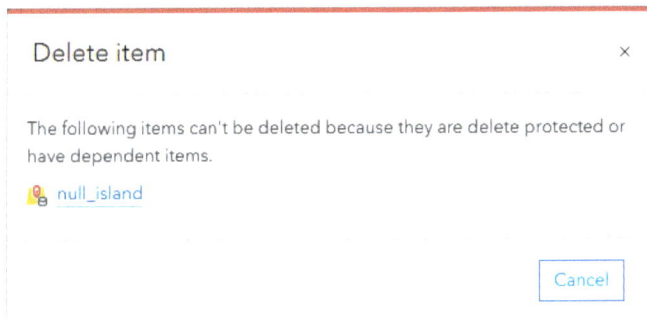

5. Perform the same workflow to delete the **null_island view** feature layer item.

Although the web map item references this view layer, ArcGIS Enterprise will allow you to delete it. The web map, however, is now broken because the view layer item was deleted.

6. Select both the **Null Island** web map item and **null_island** hosted feature layer item and then delete them.

   The **null_island** hosted feature layer still has delete protection and cannot be deleted, but this process will delete the web map item.

7. Click the name of the **null_island** hosted feature layer to open the item page and then click **Settings**.

8. Uncheck **Delete Protection** and then click **Save**.

9. Click **Delete Item** and then click **Delete** again to confirm.

10. Delete the **null_island** csv file.

Take the next step

Review the content items in your ArcGIS Enterprise organization. Ensure that the item details have been adequately completed for your business needs, that the items have the correct owners, and that they are properly shared. Mark items as authoritative or deprecated as appropriate and delete items that are no longer necessary.

Summary

In this chapter, you learned about managing content and performed the workflows to create, configure, share, and delete content items.

CHAPTER 9

Publishing ArcGIS-managed data

Objectives

- Understand data storage options.
- Identify hosted layer types.
- Understand the benefits of the ArcGIS-managed approach.
- Create a hosted feature layer in the ArcGIS Enterprise portal.

Introduction

In the previous chapter, you learned about the different types of content you can create in ArcGIS Enterprise, such as web layers, web maps, apps, data stores, and tools. In this chapter, you'll learn how to publish and store the data that is powering these items. Based on your organization's need to access, manage, and edit data, you will learn about the types of layers you can publish and how to store your data in ArcGIS Enterprise. You will learn the difference between a hosted feature layer, where the underlying data is managed by ArcGIS, and a referenced web layer, where the underlying data is user managed.

Data in ArcGIS Enterprise

Organizations store data in ArcGIS Enterprise with the purpose of having a centralized repository. Data can be accessed by other departments spread across a wider geographic space using mobile devices and web apps to connect to the data through web services. These web services are made available through a standard referred to as REST (representational state transfer). The server application acts as an intermediary between the underlying data source, which is usually a database or geodatabase, and the client application running on a browser or in a mobile device. Organizations often adopt this approach because it is faster and more efficient than having copies of local data distributed across multiple departments or network shares.

You can publish and store data in ArcGIS Enterprise in two primary ways:

- Publish hosted feature layers in which the underlying data and lifetime of the data is managed by the ArcGIS Enterprise software.
- Reference web layers in which the underlying data and its lifetime is managed by your organization administrator.

Most organizations use a mix of hosted layers and referenced layers for support. The best choice for a given web layer depends on your data strategy.

ArcGIS-managed data

ArcGIS-managed data represents a storage strategy in which ArcGIS Enterprise maintains direct control over the physical location, life cycle, and security of the datasets you choose to host. Such data becomes hosted by ArcGIS, meaning you and your organization's members interact with these layers primarily through the ArcGIS Enterprise portal and associated apps such as Map Viewer, ArcGIS Dashboards, ArcGIS Field Maps, and ArcGIS Pro. For example, once the data is shared as a hosted feature layer, ArcGIS manages the underlying database and its optimal maintenance. It relies on the ArcGIS Data Store component to automate tasks, such as database creation, maintenance, indexing, caching, and security enforcement. The members of your organization focus on maintaining the resulting web layers, such as determining the layer access level, and performing spatial analysis.

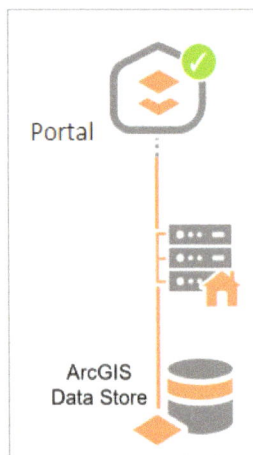

Figure 9.1. The ArcGIS Enterprise portal connects to the ArcGIS Data Store.

ArcGIS-managed data provides a clean solution, whether your team needs to produce rapid prototypes of new datasets, share information with a large number of users, or simply reduce the overhead of database administration. Once you publish layers in this manner, you can rely on ArcGIS Enterprise to handle many operational details that might otherwise require significant technical knowledge.

Benefits of ArcGIS-managed data

When organizations choose to publish hosted layers rather than referencing an external store, they often do so in pursuit of several key advantages. These benefits can include the ability to scale seamlessly, reduce complex database management concerns, and align with common workflows in ArcGIS Enterprise.

Scalability

ArcGIS-managed data can scale to accommodate everything from small departmental projects to large enterprise deployments. When data is hosted in ArcGIS Enterprise, administrators do not need to worry about intricately managing the data storage, adjusting storage parameters, or implementing complex replication strategies. ArcGIS Enterprise is designed to handle hosted layers with a large number of features without introducing unwieldy dependencies or placing too heavy a load on a team's data managers.

For example, a city government uses ArcGIS Enterprise to manage its public works data. By hosting the data in ArcGIS Enterprise, they can scale their system to support thousands of hosted layers, from small departmental projects to large citywide initiatives.

Data access

Once a dataset is published to ArcGIS Enterprise as a hosted feature layer, it becomes immediately accessible to those with appropriate permissions. This streamlines the user experience because everyone in the organization interacts with the data through familiar applications such as Map Viewer without needing direct database connections. Data creators can grant editing or viewing permissions, and the portal enforces those rules consistently.

For example, a local company publishes its asset data as hosted feature layers in ArcGIS Enterprise. This approach supports both web and field editing workflows, again without requiring direct database connections, ensuring that everyone in the organization has access to the most up-to-date information.

9

Hosted feature layer functionality

When an organization chooses to have ArcGIS Enterprise manage the data for a feature layer, it unlocks the inherent advantages of hosted feature services that ArcGIS Enterprise provides. Hosted feature layers can be edited and served out to multiple user groups or public audiences, support offline workflows for field collection, and seamlessly integrate with applications such as Field Maps or ArcGIS Survey123. This capability means that teams can collaborate on data in near real time, crowdsource updates, and enable fieldworkers to capture new information on mobile devices. Because the data resides within the ArcGIS Data Store, it benefits from automatic indexing and security controls without additional configuration or overhead.

For example, a disaster response agency uses ArcGIS Enterprise to host its emergency response data. During a crisis event, multiuser editing against these feature services allows responders to make concurrent edits. Because there isn't a mechanism to detect conflicts, the last submitted edit will be honored and made available to the entire agency. This workflow ensures that the most recent data is always available, facilitating efficient and coordinated disaster response efforts.

Data management and life cycle

When referring to ArcGIS-managed data, a key detail is how the system handles its life cycle. If a hosted layer is deleted from ArcGIS Enterprise, the corresponding service and data in the ArcGIS Data Store are also deleted. This automated approach removes the need for manual cleanup of orphaned datasets or references. Rather than juggling multiple databases or file stores, administrators can trust that any content published as a hosted item will remain strictly under the software's control for its entire lifespan. This streamlined data management model allows organizations to maintain a cleaner, more organized environment over time, without additional overhead for data removal or archiving routines.

For example, a university's GIS department uses ArcGIS Enterprise to manage its research data. When a hosted layer is no longer needed for a project, deleting it from ArcGIS Enterprise automatically removes the corresponding data from the ArcGIS Data Store, simplifying data management and ensuring a clean, organized environment.

ArcGIS Data Store

Central to the ArcGIS-managed data experience is the ArcGIS Data Store application. This software component allows you to configure data stores for your organization's hosted feature layers, hosted scene layers, and spatiotemporal feature layers in an ArcGIS Enterprise deployment. Whether installed on Linux or Windows, ArcGIS Data Store includes a setup and configuration workflow that creates a behind-the-scenes environment for storing geospatial data. Because ArcGIS Data Store is tailored to the ArcGIS Enterprise portal experience, it allows administrators to easily add, remove, and monitor the health of data store instances without custom scripting or complex database administration.

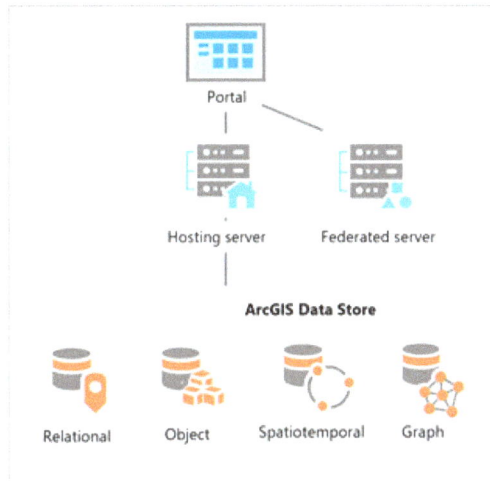

Figure 9.2. Components of ArcGIS Data Store and their relationship to the ArcGIS Enterprise portal.

ArcGIS Data Store supports multiple storage types that address various use cases:

- The **relational data store** is the data store that your ArcGIS Enterprise portal's hosting server uses to store the data for hosted feature layers. The hosting server must be configured with a relational data store before the organization administrator can add the hosting server to the portal.
- The **object data store** stores data for hosted scene layers and hosted 3D tile layers, as well as the cached query response for hosted feature layers. Starting with ArcGIS Enterprise 11.4, the object data store took the role of the now deprecated tile cache data store and stores caches for hosted scene layers. Because of this, the object data store has become part of a base ArcGIS Enterprise deployment. The ArcGIS Data Store object store requires large amounts of memory. Configure it on a machine or machines separate from other data stores.
- The **spatiotemporal big data store** supports hosted spatiotemporal feature layers and map image layers designed to store and archive high-volume, real-time observational data, including moving objects and sensor data, for efficient analysis and visualization. Primarily used with the ArcGIS GeoEvent™ Server starting from version 10.4, it can store and query significantly more data records than traditional databases. This data store also populates the location tracking service, recording the location of mobile

127

users. Because of its high demand for disk space and memory, it should be configured on separate machines from other data stores.

- The **graph store** is a database that the portal's ArcGIS Knowledge Server uses to store the entities and relationships that compose a knowledge graph. The knowledge graph services that run on the Knowledge Server site access the data in the graph store, while the object store enables efficient management of large binary objects that might otherwise be difficult to handle in a relational database.

By tailoring the data store type to the nature of the content being managed, organizations can incorporate an array of data-driven workflows without leaving the ArcGIS environment.

Web layer types that can be hosted

ArcGIS Enterprise can host many web layer types, each supporting different visualization or editing needs. Feature layers form the backbone of most hosted workflows and are stored in the relational data store. By publishing data as a hosted feature layer, users can query and edit the attributes or geometry within a web map or custom application, and the ArcGIS Enterprise portal enforces permissions, versioning rules, and other settings that the administrator configures. This approach is especially useful for collaborative editing or crowdsourcing scenarios where input from many participants is required.

Hosted feature layer views extend the concept of hosted feature layers by allowing derived layers that point to the same underlying dataset while applying specific filters, symbology, or permission rules. An organization might share a view with certain team members that omits sensitive fields or excludes certain records, ensuring that each audience sees only what is relevant to them. Because these views all refer back to the same hosted dataset, updates appear in any related view, which eliminates the need for duplicate data or complex synchronization scripts.

Hosted scene layers provide a dedicated format for displaying 3D features in the enterprise portal's scene viewer or other 3D-capable apps. Publishing a scene layer allows large multipatch or point data to be efficiently rendered on the client side, relying on server-side processing and caching to maintain performance. Administrators can thus bring detailed buildings, urban infrastructure, or terrain models into an immersive environment without imposing prohibitive loading times or GPU demands on end users.

By combining these hosted layer options with the capabilities of ArcGIS Data Store, ArcGIS Enterprise enables a highly streamlined workflow for publishing, sharing, and maintaining web layers. Organizations can rapidly create new hosted layers for a pilot project, retire or republish them as needs change, and confidently rely on ArcGIS to safeguard data integrity along the way. This holistic model allows administrators to focus on what matters: delivering the right maps, data, and analytical workflows to the right audiences, unencumbered by the intricacies of database management and infrastructure overhead.

Tutorial 9: Create a hosted feature layer

In this tutorial, you will create a hosted feature layer for field data collection. Rather than creating the layer from scratch, you will use an existing template. This saves time in creating the feature layer because this method duplicates the template properties and schema into the new feature layer. The members of your organization will then use the empty layer to create new features.

You will first log in to your ArcGIS Enterprise portal using a portal account that has, at a minimum, the creator user type assigned. You will then create the hosted feature layer using an existing template to generate the layer schema.

Create a feature layer from an existing template

1. Sign in to your organization's ArcGIS Enterprise portal.

2. On the navigation bar, click the **Content** tab. Then click **New item**.

3. In the **New item** window, click **Feature layer**.

4. Select the **Use a template** option to create an empty feature layer with fields from an existing template. Click **Next**.

5. In the **Create a feature layer** window, in the search bar, type **Crop Scouting**. Select the layer in the results.

   You will use this template to create your new layer.

   ● **Crop Scouting**

   Evaluate risk from pest infestations and disease by assessing pest pressure and crop performance.

   More details

---

6. Click **More details** to review the layer details. Then continue clicking **Next** to accept all the default layer options.

9

7. In the **New item** window, add the following descriptive item details:
   - **Title**: Field observations
   - **Tags**: field observations, hosted, Esri, Getting to Know ArcGIS Enterprise
   - **Summary**: This hosted feature layer is used to record field observations.

8. Click **Save**.

Once the layer is created, a new item is added and listed as **Feature Layer (hosted)** on the **Content** tab.

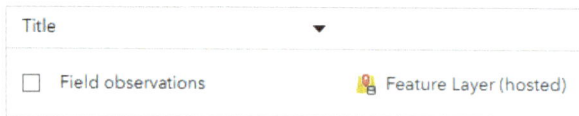

Title	▼
☐ Field observations	🗺 Feature Layer (hosted)

Make changes to a hosted feature layer

Although the layer has been generated from an existing template, modifications are often needed to ensure the layer properties match a specific industry or the parameters of the intended data collection process. Next, you will edit the layer capabilities and fields.

1. On the **Content** tab, click the **Field observations** feature layer to open the item page.

2. Click the **Settings** tab.

This opens the general capabilities associated with this layer.

3. Scroll down to the **Editing** subsection.

You will see the current capabilities enabled on the hosted feature layer as follows:
   - **Enable editing** is enabled to allow portal users to add, delete, and update features.
   - **Enable sync** is also enabled to allow portal users to make edits while disconnected from the network.

You will keep these capabilities enabled as these are useful for field data collection. At the end of the data collection process, you will want to allow your customers to download the data in various formats. For that, you will need to enable the export option.

4. Scroll to the bottom of the page. Under **Export Data**, check the box for **Allow others to export to different formats** option.

Export Data
☑ Allow others to export to different formats.

5. Click **Save**.

Next, you will add one more field necessary for the data collection.

6. Scroll to the top of the **Settings** tag and click the **Data** tab.

7. Click the **Fields** tab in the top-right corner.

All the layers fields are displayed. These fields were added from the layer template used. Now you will add one more field.

Display Name	Field Name	Type
OBJECTID	objectid	ObjectID
Observation	observation	String
Severity	severity	String
Notes	notes	String
Date	date	Date
GlobalID	globalid	GlobalID
Photos And Files	Photos And Files	Attachment

9

8. Click the **Add** button.

9. In the **Add Field** window, enter the following settings:
    - **Field Name**: Class
    - **Display Name**: Class

10. When done, click **Add New Field** to save your changes.

    The **Class** field is now displayed in the fields list, and the layer is ready for field data collection.

Summary

You reviewed at a high level the two methods of sharing data to ArcGIS Enterprise. By copying all data, a hosted feature layer is created. In this case, you are choosing the ArcGIS-managed approach to data management. On the other hand, when data is not copied and is shared by reference from a registered data store, you are choosing a user-managed approach to data management because the data is stored and maintained outside of ArcGIS Enterprise. This chapter detailed the first approach, ArcGIS-managed data. We reviewed the benefits and the types of hosted feature layers that can be published and detailed the various ArcGIS Data Store types that can be configured for use with ArcGIS Enterprise. In the next chapter, you will review the details of the user-managed approach.

CHAPTER 10

Publishing user-managed data

Objectives
- Understand data storage options.
- Review data store items.
- Understand the benefits of sharing data by reference.
- Create a data store item.

Introduction
In this chapter, we turn our attention to user-managed data in ArcGIS Enterprise. Unlike ArcGIS-managed data, which resides in ArcGIS Data Store and is fully controlled by ArcGIS, user-managed data remains in external storage, such as folders, databases, or even cloud storage. Organizations adopt this approach when they need fine-grained control over database management, security, and performance. This approach is also adopted when advanced capabilities such as versioning, attribute rules, and utility networks are used in workflows.
- **Versioning** allows multiple users to edit data simultaneously without locking the data.
- **Attribute rules** enforce data integrity through constraints and calculations.
- **Utility networks** model and analyze complex infrastructure systems.

These capabilities need to be enabled at the database level.

User-managed data
User-managed data is data that is shared by reference to ArcGIS Enterprise. In this data storage strategy, the underlying data resides outside the immediate purview of the ArcGIS Enterprise portal. Instead of copying that data to ArcGIS Enterprise during publishing, administrators choose to store, manage, and maintain it in their own infrastructure or data systems, such as a relational database. This publishing method is beneficial to larger or more

specialized organizations that rely on a centralized data strategy. If a team has optimized an enterprise geodatabase or maintains a file-based repository, they can continue using their existing setup. ArcGIS Enterprise connects to this external data to support the published web layers, allowing organizations to retain their established data management practices.

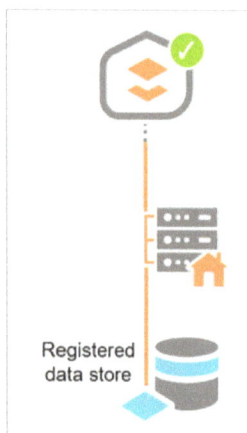

Figure 10.1. The relationship between the ArcGIS Enterprise portal and registered data store.

For many organizations, these user-managed data sources are not new: They may already be home to critical assets, project-based archives, or authoritative enterprise datasets that feed into business workflows. By configuring ArcGIS Enterprise to reference these sources directly, teams can continue to apply familiar workflows, such as database indexing, daily backups, and existing security rules. This approach stands in contrast to simply placing data under the control of ArcGIS Enterprise, which can be more convenient for certain scenarios but forfeits some of the administrative depth and advanced workflows that managers may require.

For example, a city planning department uses ArcGIS Enterprise to host its zoning and land-use data. By hosting the data, the city can scale its system to support numerous web layers without worrying about complex database management. This setup allows planners to access and update zoning information in real time using ArcGIS Field Maps, ensuring that everyone has the latest data without needing direct database connections.

In contrast, for example, a transportation agency has an extensive enterprise geodatabase containing critical infrastructure data, including road networks and traffic signals, in which editors need to make concurrent edits in an isolated environment. By configuring ArcGIS Enterprise to reference this geodatabase directly, the agency can continue using its established workflows, including multiuser editing with versioning.

Benefits of publishing data by reference

Multiuser editing and dynamic updates

One of the most significant advantages of user-managed data is the ability to accommodate multiuser editing in an isolated environment without locking or duplicating the data. Because the information remains in a relational database or other shared environment, multiple editors can simultaneously apply updates, whether they are editing attributes, geometry, or running batch processes that transform the data each night. This leads naturally to the second benefit, which is dynamic updates. Any changes that occur in the external database automatically appear when a user accesses the web layer, so there is little or no delay when it reflects the latest edits. This is because the web layer is dynamically referencing the underlying geodatabase.

For example, a county government published its parcel data by reference to ArcGIS Enterprise with the purpose of using branch versioning. By using branch versioning, each editor can create their own named version without affecting each other's changes. Once all edits are finalized, the changes are verified during the conflict review process. The vetted edits are then reconciled and posted to the default version so that the most accurate and up-to-date information is available to all users and the underlaying database.

10

Data integrity and support for complex data models

The user-managed strategy also places data integrity at the forefront. Database administrators can adopt robust measures to ensure that relationships, constraints, and rules remain consistent across all layers and tables. If advanced workflows are needed, an externally managed enterprise geodatabase is often the best environment for the sake of performance and scalability. An example of an advanced workflow is the utility network model, which requires specialized relationships among assets, such as lines, transformers, or valves.

Organizations can extend the concept of data integrity into intricate modeling of utility systems, attribute rules, topologies, or archiving for historical analysis of changes over time.

For example, a city park department wants to automate the data entry process and cut down on user errors. For that, the city publishes its data by reference to ArcGIS Enterprise to make use of on-demand attribute rules. The city creates a batch calculation rule to calculate the total number of planned trees in each park. A validation rule is also deployed to automatically flag trees in poor condition. This automated approach saves time and reduces the potential for human error, ensuring that the department's data is accurate and up-to-date.

Data management

Because the source data is stored in a database managed by the organization, the data can seamlessly be incorporated into broader data management strategies. For instance, the same database might also serve data to applications unrelated to GIS, such as business intelligence tools or asset management systems. This integration helps teams avoid silos, ensuring that their authoritative data does not end up duplicated or scattered across multiple environments. Furthermore, once a reference web layer is deleted, the associated server is automatically deleted, but the source data located in the enterprise is retained.

For example, a water utility integrates its GIS data with broader data management strategies by storing it in an enterprise geodatabase. This database serves not only GIS applications but also business intelligence tools and asset management systems. For instance, the same database provides data for hydraulic modeling, outage management, and regulatory compliance reporting. This integration prevents data silos and ensures that authoritative data is not duplicated or scattered across multiple environments. When a reference web layer is deleted, the source data in the database is retained and can be further used for hydraulic modeling.

Types of data store items

To share a web layer by reference to ArcGIS Enterprise, the external data storage location must first be registered as a data store. More specifically, the data store needs to be registered with the ArcGIS Server site that will be used to support the web service. This may be the hosting server, or it may be another federated ArcGIS Server site. Registration is the process by which ArcGIS Server is informed about how to connect to an external data source. On a foundational level, a few common storage types can be used in a user-managed scenario.

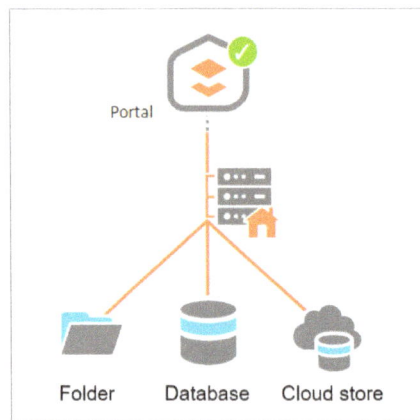

Figure 10.2. The relationship between the ArcGIS Enterprise portal and storage types.

One of the most common types of user-managed stores is the enterprise geodatabase. These **databases** are set up in relational database management systems, such as Oracle, Microsoft SQL Server, PostgreSQL, or other supported platforms. Within the geodatabase, organizations may implement versioning, role-based security, sophisticated indexing, or topological relationships for geometry validations.

Alternatively, some smaller teams may rely on folders containing file geodatabases or mobile geodatabases for simpler local editing.

ArcGIS Server can also connect to cloud stores if they have been appropriately registered, which might involve referencing Amazon S3 or similar storage services that hold large volumes of geospatial files. Additionally, organizations can use cloud data warehouses, such as Google BigQuery, Snowflake, or Amazon Redshift, to store and manage vast amounts of geospatial data, providing scalable and flexible solutions for data-intensive applications.

Web layers that can reference a registered data store

When a data store is registered with ArcGIS Server, ArcGIS Enterprise can create several types of web layers, each serving a different set of needs. These references ensure that ArcGIS Enterprise only points to the external data rather than copying it, preserving the advanced capabilities and structure of the original environment.

- A **feature layer** delivers geometry and attribute information to clients as discrete features, which can be queried, displayed, and edited if permissions allow. For example, city planners might rely on a feature layer that references an enterprise geodatabase storing parcel boundaries. Because the layer references the external database, any updates performed directly on the geodatabase become visible in the application as soon as they are committed.
- **Scene layers** are specialized for 3D contexts, allowing the rendering of multipatch geometry, 3D buildings, or point cloud data in a web scene. When scene layers reference an external data store, large or complex 3D datasets remain in their optimized location and benefit from any back-end infrastructure that has been established for high-volume storage.
- **Imagery layers** expose large raster or mosaic datasets in a dynamic or cached fashion. By pointing to user-managed data, ArcGIS Server can rely on that external source for performance optimizations, advanced image parameters, and immediate updates if new images are swapped into the mosaic. This approach is often critical for organizations, such as environmental agencies or agricultural companies, which generate or process new imagery regularly.
- **Map image layers** dynamically render map content on the server side, supporting advanced cartographic output or labeling logic that can be crucial for professional-grade mapping. Through user-managed data, organizations gain the benefit of full

control over symbology, labeling rules, and data transformations while letting ArcGIS Server handle final rendering.

- **Vector tile layers** deliver vector-based data in a tiled format. Although some teams prefer to generate hosted vector tile packages using ArcGIS Data Store, many large organizations already maintain massive point, line, and polygon datasets in external databases and simply want to tile them in place for efficient consumption. By registering their data store, they can produce vector tiles from the user-managed source, giving them the performance benefits of a tiled approach and the flexibility of an external environment.

By understanding the various ways that ArcGIS Enterprise can point to external data sources, organizations can tailor their publishing approach to match each project's needs. Some workflows might favor a feature layer for straightforward editing, whereas others might use scene layers for immersive 3D visualization. In each scenario, the key principle remains the same: The data is never copied or hosted but instead references the external data store. This ensures that the organization retains elevated control over security policies, update schedules, and advanced geodatabase functionalities, all while harnessing the powerful capabilities of ArcGIS Enterprise for web-based access and collaboration.

Tutorial 10: Create a data store in ArcGIS Enterprise

In this tutorial, you will create a new data store item in the ArcGIS Enterprise portal. Then you will use that data store item to publish multiple web layers. In the next chapter, we will cover one of the most common approaches to publishing data to ArcGIS Enterprise using ArcGIS Pro.

Create a data store item

You will use a geodatabase to create a data store item and then publish a web layer from it. For this tutorial, you will need access to an .sde file that stores the database connection for the enterprise database you want to create the data store item for.

1. Sign in to your organization ArcGIS Enterprise portal. Ensure that your account has publishing privileges.

2. On the navigation bar, click the **Content** tab and then click the **New item** button.

3. In the **New item** window, scroll down and select **Data store**.

4. For **Select the type of data store to add**, select **Database** and then click **Next**.

5. For **Select the type of database**, select **Relational database** and then click **Next**.

6. For **Specify connection or location information to allow the ArcGIS Server sites to access the data**, click **Select File**.

7. Browse in File Explorer to the location of the .sde file that stores the database connection to the enterprise geodatabase you want to create a data store item for.

8. Select the .sde file and click **Open**.

   The connection information appears. Your connection information will be different, based on your server, instance, database, and other properties.

9. Click **Next**.

10. Check the box next to your server to select it.

   Select the ArcGIS Server sites to which you want to add your data store.

   The site must have access to the connection or location you specified in the previous step.

Server	Status
☐	
☑ https:// **your server URL** (Hosting Server)	● Normal

11. Click **Next**.

12. For the **Data store connection** settings, enter the following information:
   - **Title**: Your data store item name.
   - **Folder**: An appropriate folder to save your data store in.
   - **Tags**: Any appropriate tags for your data store.
   - **Summary**: This is a new data store item created directly in the ArcGIS Enterprise portal.

13. Click **Create connection**.

   On the **Content** tab, the item is listed as a data store.

☐ DataStore	🖳 Data store (database)

Next, you will use the data store item to publish multiple web feature layers directly from the ArcGIS Enterprise portal.

Publish multiple web feature layers

In these steps, you will bulk publish web feature layers from the data store item. This will eliminate the need to manually publish each web layer one by one.

1. On the **Content** tab, click the data store item you previously connected to open the item page.

2. Click the **Layers** tab and then click **Create Layers**.

3. For **Choose Layer Properties**, change the **Time zone of the data** to the time zone you are located in. Then click **Next**.

4. For **Choose folder**, select the same folder you created the data store item in. Click **Next**.

5. For **Choose server**, verify that your server is selected and click **Start publishing**.

   Depending on how many feature classes your enterprise geodatabase has, you will see a feature layer and a map image layer for each feature class. In ArcGIS Enterprise, map image layers and feature layers are often published together because they serve complementary purposes. Map image layers are typically used for displaying data dynamically or as cached image tiles and are great for faster rendering and advanced symbology. Feature layers provide access to the underlying geographic features and their attributes, which are needed for editing and running analysis.

Summary

In this chapter, you have explored the advantages of publishing to ArcGIS Enterprise data that references a registered data store. You examined the various types of storage locations that can be registered as a data store, along with their corresponding web layer types. Additionally, through the guided tutorial, you created a data store item that references an enterprise geodatabase. You used the bulk-publishing feature in ArcGIS Enterprise to publish multiple layers simultaneously. The upcoming sections of the book will provide a more in-depth examination of all the elements discussed so far by following a structured workflow with an overarching example in the next five chapters.

PART 3

Introduction to common workflows in ArcGIS Enterprise

THE NEXT FIVE CHAPTERS COVER ESSENTIAL WORKFLOWS FOR publishing, editing, analyzing, and collaborating in ArcGIS Enterprise. Building on previous chapters, you'll follow five guided tutorials that cover a fictional use case scenario. In these tutorials, you will learn how to connect to an enterprise geodatabase, publish a web feature layer, and edit data over the services. You will also learn how to perform a spatial analysis, share the results of your findings in a dashboard, and collaborate with other organizations to promote data accessibility. By following these workflows, you'll gain practical experience using sample data, helping your organization keep its data current and make informed, data-driven decisions.

CHAPTER 11

Publishing data to ArcGIS Enterprise

Objectives

- Create a database connection in ArcGIS Pro.
- Prepare data for publishing.
- Connect to ArcGIS Enterprise from ArcGIS Pro.
- Publish by reference a web feature service to ArcGIS Enterprise.

Introduction

Publishing data by reference in ArcGIS Enterprise allows organizations to efficiently manage and share geospatial information while maintaining data integrity between the source geodatabase and the web feature service. This approach enables users to publish web feature layers directly from an enterprise geodatabase, ensuring that the data remains centralized and authoritative. In this section, we will guide you through the essential steps of signing in to your ArcGIS Enterprise portal through ArcGIS Pro, connecting to the enterprise geodatabase, and preparing your data for publication. By publishing a web feature layer that references a registered data store, you can streamline access to your geospatial resources and enhance the use of your data across various applications.

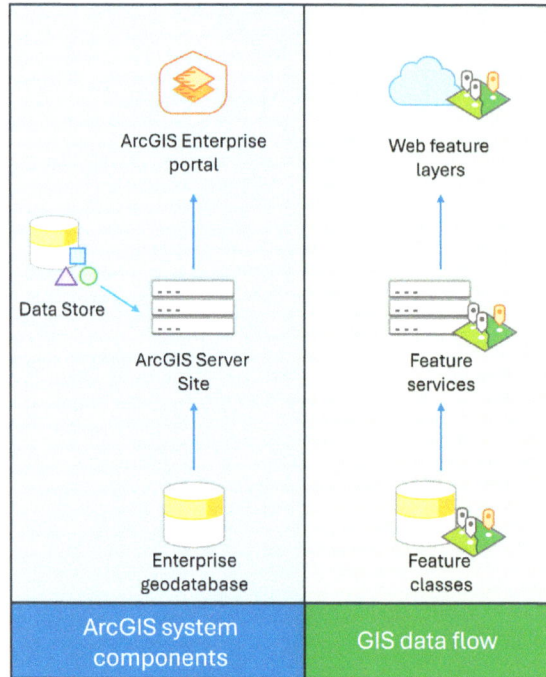

Figure 11.1. Publishing workflow diagram.

Tutorial 11: Publish a web feature service from ArcGIS Pro

In this tutorial, you will assume the role of a GIS specialist for Medio County. In this fictional scenario, the county's GIS department is transitioning from direct database editing to web-based editing within ArcGIS Enterprise. The purpose of this transition is to facilitate an authoritative data flow between the source geodatabase and the clients accessing that data through web feature services. Your first task in this project is to share the department's data as a web feature layer to ArcGIS Enterprise.

To complete your task, you'll establish a connection to the enterprise geodatabase, which serves as the primary data repository for the department. To prepare the data for sharing, you will enable geodatabase capabilities, including global IDs and editor tracking. Then you will publish a web feature service that references the data in the enterprise geodatabase.

Connect to an enterprise geodatabase

You will connect to the department enterprise geodatabase as a database user that has privileges to load data.

1. Open ArcGIS Pro. Under **New Project**, click **Map**. Save your map as **Medio Project**.

2. In the upper-right corner of the screen, click the profile picture icon to sign in. Click **Manage Portals**.

3. On the **Portals** page, click **Add Portal**.

    Depending on the portals, if any, that you are connected to, you may see other portals listed. If you already see the portal listed, you can skip the step to add the portal and proceed to making it the active portal.

4. In the **Add Portal** dialog box, type the URL for your portal and click **OK**.

    > *Tip: For example, the URL format would be https://gis.example.com/portal/.*

    The portal you added appears in the list.

5. In the list of portals, right-click the URL and click **Set As Active Portal** to make the new portal connection your active portal.

    You have added the portal and set it as the active portal. Next, you will sign in to it.

6. Right-click the portal you just added again and click **Sign in**.

    Note: To successfully complete this tutorial, make sure you are connecting to a portal account that has publishing capabilities.

7. After you successfully sign in, click the back arrow to return to the project.

    In the project's upper-right corner, the name of the portal you are connected to is listed.

8. In the **Catalog** pane, right-click the **Database** folder and click **New Database Connection**.

11

9. In the **Database Connection** window, enter the parameter values as follows:
   - For **Database Platform**, choose the type of relational database where your data is stored. This workflow will use **SQL Server**.
   - For **Instance**, type the name of your database instance. This workflow will use MCountySQL.
   - For **Authentication Type**, select **Database authentication**.
   - For **User Name**, type the name of the database user with sufficient privileges to load data into the database.
   - For **Password**, type the password associated with the account.
   - For **Save User/Password**, keep the box checked.
   - For **Database**, select the database you want to work with from the drop-down menu. This workflow uses **MedioDB**.

Note: If you use a relational database management system (RDBMS) other than SQL Server, some of the parameter values might differ. Moreover, some values will be specific to the database instance you are targeting (such as instance name and user credentials).

10. Click **OK**.

The database connection appears in the Catalog pane as **SQLServer-MCountySQL-MedioDB(gis).sde**. This enterprise geodatabase will act as the primary geodatabase of the department. You will see that the naming convention for the enterprise database connections is DatabasePlatform-Instance-DatabaseName(username).sde. Whenever you make a connection, the name will have this convention.

11. Click the arrow next to the name to expand the newly created database connection.

In this tutorial, the database is empty. You will use this database to load data into the enterprise geodatabase you connected to earlier.

12. In a browser, navigate to **links.esri.com/GTKEnterpriseData**. On the item page, click **Download**.

The dataset is downloaded as a copy on your local machine. You will unzip the file on your local machine and then add it as a folder connection to your ArcGIS Pro project.

13. Unzip the data folder.

14. Back to ArcGIS Pro, in the **Catalog** pane, right-click the **Folders** folder and then click **Add Folder Connection**.

11

15. Navigate to the folder where the downloaded file geodatabase is located and then click **OK**.

The folder is now listed in the **Catalog** pane, under active folders.

16. Expand the file geodatabase. It contains three feature classes:
- **City_Boundaries**: A polygon feature class that represents the city boundaries in Medio County.
- **Road_Centerlines**: A polyline feature class that represents all major and local roads in the city.
- **Schools**: A point feature class that represents the locations of all the schools in the city.

17. Press the **Ctrl** key and select all three feature classes.

18. With the selection active, drag the three feature classes and drop them into your enterprise geodatabase.

19. When the operation finishes, right-click the database connection and click **Refresh**.

   Notice how the database username is included in the feature class names. This makes it easier to indicate the data owner of these feature classes.

20. Next, you will prepare the data for publishing.

Prepare data for publishing

In this section, you will get familiar with the data. You will also add Global IDs and enable editor tracking to facilitate the web-editing experience on the published data.

1. In the **Catalog** pane, within the database connection, right-click the first feature class, **City_Boundaries**, and then click **Manage**.

2. On the **Manage** tab, check the following capabilities:
   - Enable **Global IDs** as a prerequisite for the publishing process, which is necessary for maintaining object uniqueness.
   - Enable **Editor tracking** to maintain a record of the editor who created or modified the data and a time stamp of when the edit occurred. Many organizations find editor tracking helpful to maintain accountability and transparency.
   - Enable **Archiving** as a prerequisite to enabling the syncing capability on the feature layer. This allows an editor to take the data offline, make edits in the field, and then synchronize the edits once they are connected to the internet again.
   - Enable **Replica Tracking** as a prerequisite needed when sharing your data with an ArcGIS Online organization using distributed collaboration.

3. When done, click **OK**.

   Manage geodatabase functionality
   ☐ Versioning
   ☑ Archiving
   ☑ Replica Tracking
   ☐ Attachments
   ☑ Global IDs
   ☑ Editor tracking

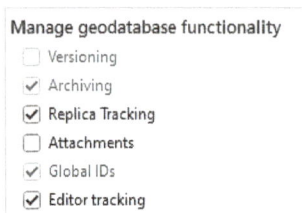

4. Repeat the previous steps to prepare the other two feature classes for publishing.

5. Multiselect all three feature classes from the enterprise geodatabase and then click and drag them to the map.

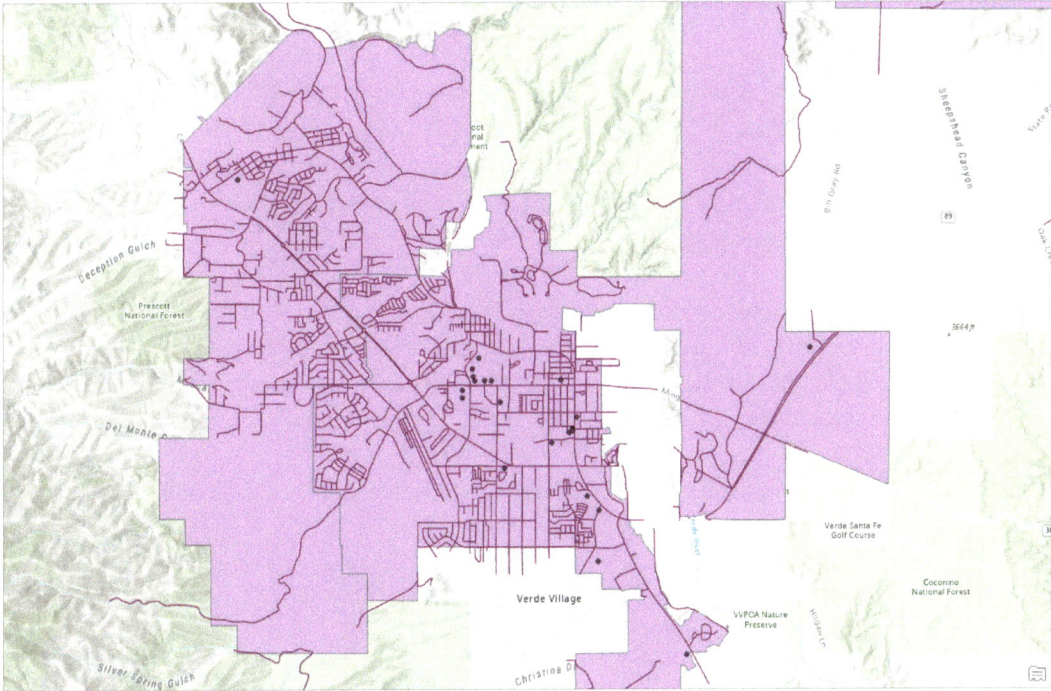

Now that the data is prepared and added to the map, the next step is to publish the department's data to the ArcGIS Enterprise portal. To do that, you will first connect to your organization portal.

Publish data to the portal

In this final part of the tutorial, you will connect to your organization portal from ArcGIS Pro. You will use the **Share As Web Layer** tool to publish the feature classes from your enterprise geodatabase as a web feature layer.

1. On the **Quick Access** toolbar at the top, click the **Save Project** button.

Next, you will publish the data.

149

2. On the ribbon, click the **Share** tab. In the **Share As** group, click **Web Layer**.

   The **Share As Web Layer** pane appears. Here, you can enter the parameters for the web layer and analyze it for errors before publishing.

3. In the **Share As Web Layer** pane, enter the following information:
   • **Name**: Medio Public Assets.
   • **Summary**: This web layer represents the legacy data of Medio County, GIS Department. It includes the city boundaries, road centerlines, and school locations.
   • **Tags**: public assets, data management, Medio County, Esri.

   Next, you will complete the **Data and Layer Type** information. To share data that references registered data, a map image layer is automatically included. To support feature querying, visualization, and editing, you must also enable the **Feature** option. This will create a web feature layer and a map image layer in your portal.

4. For **Data and Layer Type**, under **Reference registered data**, check the box for **Feature**.

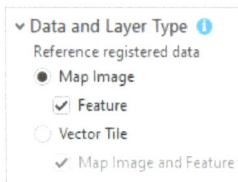

5. For **Location**, under **Portal Folder**, click the drop-down menu and select **Create new folder**. For the folder's name, type **County Assets**.

   You will use this folder to store all the department's layers.

6. For **Sharing Level**, click **Organization**.

   Checking your enterprise organization ensures all members of your organization will have access to this web layer. Next, you will analyze the web layer to check for errors.

7. At the top of the pane, click the **Configuration** tab.

8. Under **Layer(s)**, next to **Feature**, click the **Configure Web Layer Properties** (pencil icon).

9. For **Operations**, check the box for **Enable Sync**.

   This allows editors to take the data offline.

10. At the bottom of the pane, for **Sync**, change the **Version Creation** to **None**.

    This is because your data is not versioned, so all the edits will be synchronized directly with the database.

11. Under **Finish Sharing**, click **Analyze**.

    Several errors and warnings are returned. You must address all errors before publishing, but you can leave the warnings.

12. Expand the error regarding the data source being registered with the server.

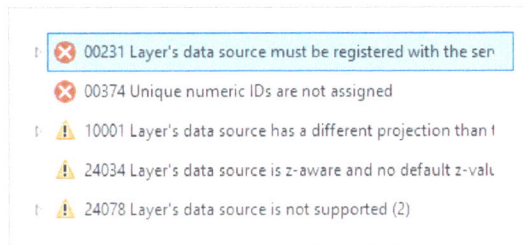

The first error indicates the layer data source is not registered with the server. You are publishing three feature layers, so there are three errors. To address these errors, you will register the MedioDB enterprise geodatabase with the ArcGIS Server site by creating a data store item.

Note: When you publish web services to ArcGIS Enterprise and choose to reference registered data, the data source must be registered with ArcGIS Server. This registration allows the server to access your data and use it as the source for web layers. Creating a data store is the key to making your data accessible to the server. A data store can be any location—enterprise database, folder, cloud store, or NoSQL database—that houses the data you want to use. After registering the data with the server, the published web service establishes a direct connection to the data source. This connection ensures that the web services reference the data in the data store without duplicating it.

---

13. Expand the first error and right-click the first one. Click **Register Data Source With Server**.

14. In the **Add Data Store** window, provide the connection details for the data store:
  • **Title**: MedioDB_DataStore
  • **Tags**: public **assets**, **data management, Medio County, Esri**
  • **Portal Folder**: County Assets
  • **Sharing Level**: ArcGIS Enterprise

15. Click the **Validate** button to validate the server database connection.

16. Click **Create**.

A check mark appears in front of the first server message, which indicates the layer's data source is registered with the server.

Although it appears that you must register the other two layers, adding the data store one time will correct the issue for all layers with the same error because all the data is stored in the same geodatabase. When you analyze the web layer again, those errors will be resolved.

17. Click **Analyze**.

The errors regarding registering the data with the server are resolved. Finally, you will clear the last error by assigning unique numeric IDs. Assignment of unique IDs is a requirement when sharing data as a web layer. It ensures layer IDs remain static when the web layer or service is overwritten.

18. Right-click the **Unique numeric IDs are not assigned** error message and click **Auto-Assign IDs Sequentially**.

The error is resolved. You will now publish the web layer.

19. Click **Publish**.

After the publishing process is complete, at the bottom of the pane, a message confirms the web layers have been successfully shared. The message also contains a link to manage the web layers in your ArcGIS Enterprise portal. You will use this link to access the web layers directly in the ArcGIS Enterprise portal.

20. Click the **Manage the web layer** link.

If necessary, in the upper-right corner of the page, sign in with the same portal account to access the published data.

The item page for the web layer you published appears on a browser tab.

21. On the navigation bar, click the **Content** tab. Navigate to the **County Assets** folder and view its contents.

The **Public Assets** folder contains three portal items that were created when you published the **Medio Public Assets** web layer:

- A data store item that ensures the ArcGIS Server site has access to the published data.
- A map image layer that is available only if you are sharing to an ArcGIS Enterprise portal. It is automatically created when you publish data by referencing a registered data store.
- A feature layer that supports vector querying, visualization, and editing.

Summary

In this chapter, you assumed the role of a GIS specialist for Medio County, successfully completing the initial task of the project. You shared the county's data as a web feature service to ArcGIS Enterprise, supporting the department's transition from direct database editing to a web-based editing approach. As a data owner, you established a database connection to the department's enterprise geodatabase and used that connection to import data from a file geodatabase. After the data was integrated, you enhanced the feature classes by adding global IDs and enabling editor tracking. Finally, you published the legacy data from ArcGIS Pro by referencing data from the enterprise geodatabase. In the next chapter, you will explore the web-based editing experience using both ArcGIS Enterprise and ArcGIS Pro.

CHAPTER 12

Editing against feature services

Objectives

- Make edits to a web layer in Map Viewer.
- Add a web feature layer to ArcGIS Pro.
- Make edits to the underlying enterprise geodatabase.
- Assess web edits being referenced to the enterprise geodatabase.

Introduction

When a web layer is shared, a service becomes available and is made accessible through that layer. These web layers can be used for visualization or editing purposes across the ArcGIS platform, which includes ArcGIS Pro, Map Viewer, and Scene Viewer. This web-based editing capability enables a broader audience to contribute to and enhance your data. It may include fieldworkers, analysts from various departments within your organization, or even crowd-sourcing volunteers who can provide significant input to your data through an intuitive editing interface, all simultaneously. In the previous chapter, we demonstrated how to publish data by referencing an enterprise geodatabase. In this chapter, we will focus on making edits to the web layer using various ArcGIS clients. We will then assess how these changes are automatically reflected with the data source, the enterprise geodatabase.

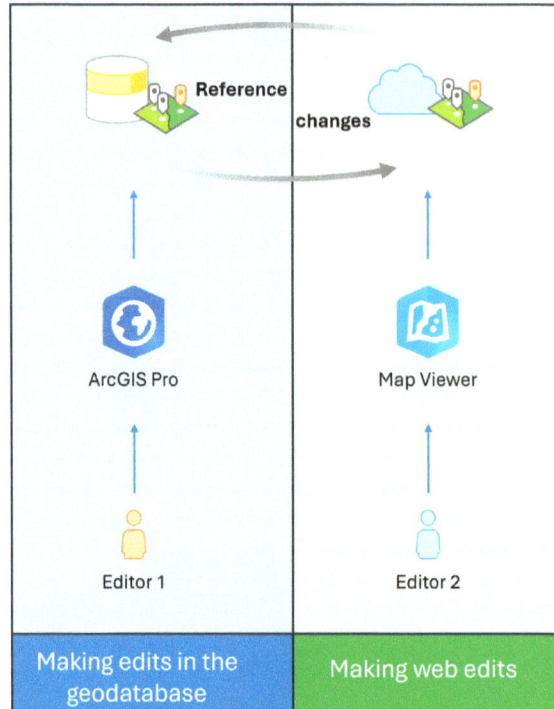

Figure 1.1. Web-editing diagram.

Tutorial 12: Practice multiuser editing over the web

In this tutorial, you will continue your work as a GIS specialist for Medio County, which is currently in the process of enhancing the county's GIS department. The department is moving from direct database editing to web-based editing within ArcGIS Enterprise, thereby enabling a reliable data flow between the source geodatabase and the clients who access that data through web feature services. You will evaluate the web-editing experience before its implementation for frontline workers.

Your next set of tasks is to add the Medio County web feature layer into Map Viewer, modify attributes and geometry of existing features, and create new features. You will then return to ArcGIS Pro to verify that the changes made online are represented in the enterprise geodatabase. Finally, you will add the web feature layer to ArcGIS Pro and perform further modifications.

Add web layer to Map Viewer

You will connect to your organization portal account and sign in with a portal user that has editing privileges. You will open Map Viewer, add the web feature layer you published in the previous chapter, and make geometry and attribute edits to the features.

1. In the browser, type your ArcGIS Enterprise organization URL to access the portal website.

   > *Tip: The URL format is https://gis.example.com/portal/home/.*

2. Sign in as a portal user that has at least editing credentials.

   This ensures you will be able to make edits to the web feature layer.

3. Next to your profile picture, click the **App Launcher** (nine-dots icon). From the apps list, click **Map Viewer** to open the app on a new tab.

   Map Viewer appears. The visual representation of your map might be different based on your account or organizational configurations. It may depict the United States, a global view, or another specific area. The only layer present on the map is the basemap, which offers essential geographic context, including water bodies and political boundaries. The default basemap is **Topographic**; however, your map may use a different basemap based on your organization's preferences.

   Map Viewer includes two vertical toolbars—the **Contents** (dark) toolbar and the **Settings** (light) toolbar.

   Next, you will add the web feature layer you published earlier to the map.

4. On the left of the screen on the **Contents** toolbar, if necessary, click **Layers** to display the **Layers** pane.

5. In the **Layers** pane, click **Add**.

   > *Tip: You can use keyboard shortcuts to access common workflows in Map Viewer, such as adding a map layer and opening the panes to style and filter layers. To view the full list of keyboard shortcuts in Map Viewer, press Alt+? on Microsoft Windows or Option+? on Mac.*

12

6. In the **Add layer** pane, ensure you are searching for layers from **My organization**. If it is not set to that, click the drop-down list and change the location.

7. In the search bar, type **Medio Public Assets**.

8. From the search results, add the **Medio Public Assets** feature layer to the map.

> *Tip: Ensure you are adding the feature layer and not the map image layer. This ensures you can make edits to the web layer.*

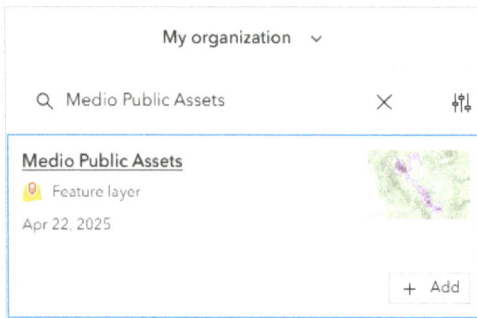

The layer is added to the map and the map zooms to its extent.

Note: If you can't see the web layer of interest, ensure the item is shared with the entire organization, or ensure that the portal account you are connected to has access to the layer item.

The Medio Public Assets layer is now listed as an active layer on the map.

Make web edits

You will now proceed to evaluate the editing experience. A newly established school has commenced operations. To commemorate this development, you are required to identify the location of the New Visions Academy on the map and update the **Status** attribute to indicate that it is open. County officials have constructed a new road directly in front of the school, so you will also need to create a new road feature.

1. In the **Layers** pane, click the left back arrow to view the list of layers.

2. Click the arrow next to **Medio Public Assets** to expand the group of layers.

3. Click the **Schools** layer.

   This allows you to access that individual layer's options on the toolbar.

4. From the **Settings** toolbar, click the **Filter** tab.

   You will use the **Filter** tool to search for a specific school feature.

   > Tip: Notice at the top of the Filter pane that Schools is the active layer. If you want to filter another layer from the map, use the right arrow next to Schools to change the active layer.

5. In the **Filter** pane, click **Add new**.

6. Under **Condition**, apply the following parameters:
   - **Field name**: SchoolName
   - **Expression operator**: is
   - **Value**: New Visions Academy

Condition	...
SchoolName	⌄
is	⌄
New Visions Academy	⌄

12

The map filters based on the expression you entered. You will notice only one school feature is now displayed.

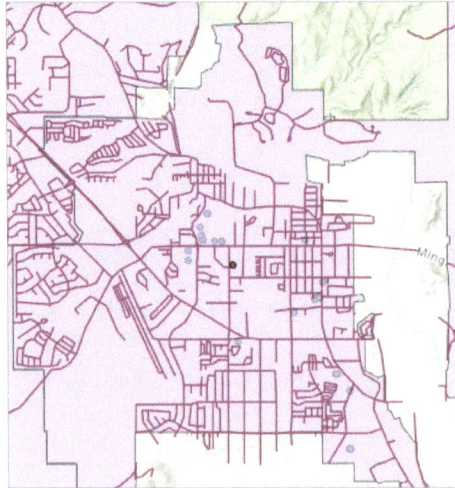

7. On the **Settings** toolbar, click the **Edit** tab.

    The **Editor** pane allows you to make edits to existing features and create new ones. In this example, you will make an attribute change to the **New Visions Academy** feature.

8. In the **Editor** pane, under **Edit features**, click **Select**.

9. With the **Select** feature active, zoom in and click the New Visions Academy point feature on the map.

10. If there is more than one feature in that location, under the **Schools** section, select **OBJECTID: 30**.

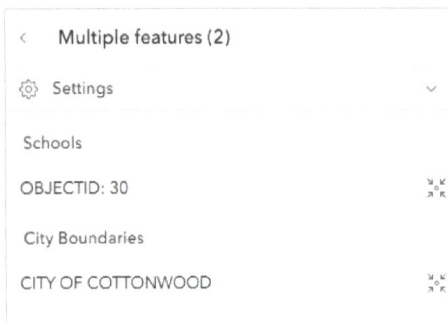

A form appears that allows you to update the feature's attributes.

11. Scroll down until you find the **Status** field. Click the field and type **Open**.

12. Click **Update** to apply your change.

The final edit you will make is to create a new road line that will symbolize direct access from the main road to the school.

13. Open the **Filter** pane. Click **Clear all** and then close the pane.

This removes the filter, and all features are now displayed on the map.

14. On the map, zoom in to the **New Visions Academy** point feature.

You will create a new road line that connects the main road line, 6th Street, to the school entrance.

15. Open the **Editor** pane. Under **Create features**, click **Road_Centerlines**.

The editing cursor activates, which will allow you to create a new road feature.

16. On the map, click to start drawing a road feature that surrounds the school feature. Use the basemap as a reference for the road.

17. When done, double-click to complete the drawing.

12

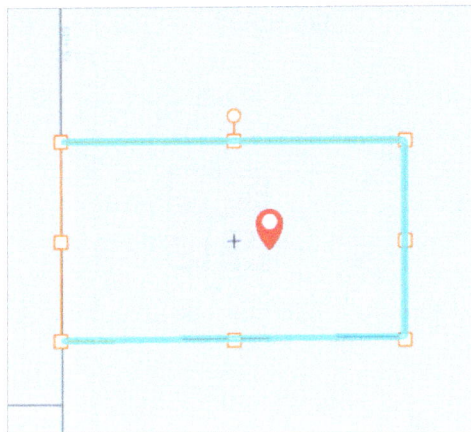

In the **Create features** pane, a form **with all the fields for the Road_Centerlines layer appears.**

18. **For the STREET NAME and FULL NAME fields, type** New Vision Road.

19. When done, click **Create.**

The new road feature is now created.

Assess the web edits in the geodatabase

In this session, you will open the ArcGIS Pro project and navigate to the enterprise geo-database that is the data source for your web feature layer. There you can confirm that the edits you have just made online are also available in the geodatabase. You will use the editor tracking fields to assess which member of the organization made the change and the time and date of the change.

1. Open the **Medio Project** in ArcGIS Pro.

2. In the **Catalog** pane, confirm you have an existing database connection to the MedioDB enterprise geodatabase.

   Tip: If you need to create a new database connection, follow the steps from the chapter 11 tutorial from the "Connect to an Enterprise Geodatabase" section.

3. If necessary, expand the database connection and drag all three feature classes to the map.

4. In the **Contents** pane, right-click **Map** and click **Properties**.

5. In the **Map Properties** window, on the **General** tab, rename the map **Database Connection** and then click **OK.**

   You will use this map to connect to the database to assess the edits that were made online. Next, you will search for the New Visions Academy.

6. On the ribbon, click the **Map** tab. Under the **Selection** group, click **Select By Attributes**.

7. In the **Select By Attributes** tool, enter the following parameters:
   - **Input Rows**: Schools
   - **Expression**: Where SchoolName is equal to New Visions Academy

8. Click **Apply** and close the pane.

9. On the map, zoom into the area of the highlighted feature.

10. On the **Map** tab, in the **Selection** group, click **Attributes**.

11. In the **Attributes** pane, scroll down to the **Status** field and ensure the new attribute is **Open**.

   This is the change you made previously to the web layer in Map Viewer.

12. Scroll down to the end of the pane, where you will see four fields:
   - **created_user:** Records the user who created the feature
   - **created_date:** Records the date and time the feature was created
   - **last_edited_user:** Records the user who last edited the feature
   - **last_edited_date:** Records the date and time the feature was last edited

   In the previous tutorial, you enabled editing tracking to maintain a record of the editor who created or modified the data and a time stamp of when the edit occurred. The information is stored in these four system-maintained fields.

For the **New Visions Academy** feature, the **created_user** and **created_date** fields are empty. This is because editor tracking was not enabled when the feature was created, so the information was not recorded. However, if you check the **last_edited_user** and **last_edited_date**, you will see the portal account you used to make the edit and a time stamp for when the edit was made.

Now check the new road line you created.

13. In the **Selection** group, click **Select**. With the **Select** tool enabled, click the **New Vision Road** line on the map.

Notice the **Attributes** pane changes to display the attributes of the selected feature. This is the new line feature you created in Map Viewer.

14. Scroll down until you find the editor tracking fields.

For the road feature, you will notice that all four editor tracking fields are populated. The **created_date** represents the time stamp of the feature creation, and the **last_edited_date** represents the moment when you added the street name and the full name for this feature. The **created_user** and the **last_edited_user** values are the same because the same portal user created and edited the feature.

Now let's see how these fields change when you make a new edit from the database connection.

15. In the **Attributes** pane, change the **ST_Category** field to 2.

16. Assess the editor tracking fields again.

You will notice the **last_edited_user** and **last_edited_date** field values changed to record the new edit you made. In this case, the **last_edited_user** is not the portal user but rather the database user you made the database connection with. In this example, this GIS user is also the data owner.

Make edits to a web layer in ArcGIS Pro

Map Viewer is not the only client application that allows you to consume and make edits to a web layer. In this final part of the tutorial, you will add the **Medio Public Assets** web layer to ArcGIS Pro.

1. In ArcGIS Pro, ensure you are connected to your portal organization as a user that has access to the **Medio Public Assets** web layer.

2. On the ribbon, go to the **Insert** tab. In the **Project** group, click the **New Map** option.

   This creates a new map in your ArcGIS Pro project.

3. Rename the map **Feature Service** map.

4. At the top of the **Catalog** pane, click the **Portal** tab.

   The tab allows you to access and browse content from your active ArcGIS Online or ArcGIS Enterprise portal.

5. Click the **My Organization** cloud icon.

   This displays all the content shared with your organization.

6. Locate the **Medio Public Assets** web feature layer and drag it to the active map.

   The layers are also added to the **Contents** pane.

7. Examine how the layers in the **Contents** pane are displayed in both the **Feature Service** map and the **Database Connection** map.

   There isn't really much difference between these two maps. If you are not the person who published this data, or are unfamiliar with it, you might want to have more information to know where this data is coming from.

8. To check the data source of layers in a map, in the **Contents** pane, click the **List By Data Source** tab to list layers by their data source.

   Displaying layers by their data source is a quick method to identify whether a layer in a map is pointing to an SDE database connection or to a feature service.

Summary

In this chapter, you continued your work as a GIS specialist, incorporating the web feature layer into Map Viewer, editing the existing School feature layer, and creating a new road feature. You evaluated the changes in ArcGIS Pro through the database connection to confirm that the web edits are referenced back to the source geodatabase. By using editor tracking, you identified the editor responsible for the most recent changes to the data. Lastly, you integrated the same web feature layer into ArcGIS Pro and explored methods to recognize where data resides based on the data source. In the upcoming chapter, you will advance your work by conducting a spatial analysis of the Medio County data.

CHAPTER 13

Performing spatial analysis

Objectives

- Create a new web map.
- Add hosted feature layers from ArcGIS Online.
- Use analysis tools such as Filter by Attributes, Summarize Nearby, and Find Closest.
- Save analysis results in a web map.

Introduction

Spatial analysis is a fundamental aspect of GIS. It enables organizations to identify patterns, relationships, and trends that inform decision-making and strategic planning by performing spatial analysis. Various GIS products, such as ArcGIS Pro, ArcGIS Online, and ArcGIS Enterprise, offer tools for performing spatial analysis. These tools allow users to conduct a wide range of analyses, from simple attribute filtering to complex spatial modeling.

For example, urban planners can perform suitability analysis to identify optimal sites for planting canopy trees by analyzing population density, proximity to existing services, and land use. Health departments use spatial analysis to track disease outbreaks and plan interventions. By mapping the spread of diseases and analyzing demographic data, planners can identify hot spots and allocate resources more effectively. Retail businesses perform market analysis by analyzing demographic data, competitor locations, and traffic patterns to find the best locations for a new store.

As a platform for spatial analysis, ArcGIS Enterprise integrates seamlessly with other ArcGIS products and data sources, such as the ArcGIS Living Atlas of the World, to enhance the depth and accuracy of analysis. One of the key benefits of performing spatial analysis in ArcGIS Enterprise is the ability to manage and analyze large datasets in a centralized, secure environment. This capability ensures data consistency and integrity across the organization. Additionally, the results of spatial analysis can be saved as hosted feature layers and shared through web maps, facilitating collaboration across teams.

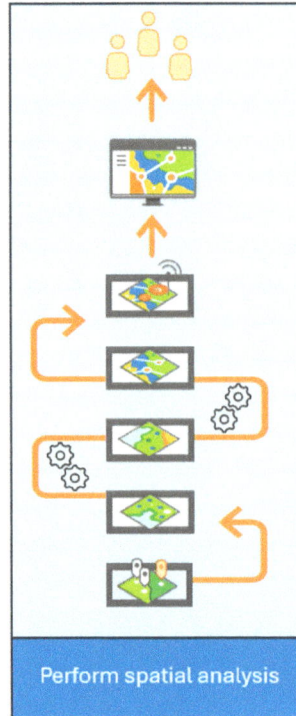

Figure 13.1. Spatial analysis diagram.

Tutorial 13: Perform a school accessibility analysis

In this tutorial, you have been assigned a new task to assess school accessibility in Medio County. Specifically, you will examine road conditions near schools, identifying unpaved roads within a half-mile radius. These areas will become the county's main priority for development. Additionally, you will identify school areas that lack a nearby public library, providing the county with insights on how to allocate funds and plan for new library facilities to ensure student access to necessary resources. Because the county's data is already published to ArcGIS Enterprise, you will use Map Viewer to perform the spatial analysis.

Find unpaved roads in school proximity

First, you will use ArcGIS Enterprise and Map Viewer to find unpaved roads that are within a half-mile radius of Medio Middle School. Then using Map Viewer, you will add the Medio Public Access web feature layer and calculate the miles of unpaved road segments within that radius. With this information, you can then prioritize roads to be fixed first because they are near the school.

1. In the browser, type your ArcGIS Enterprise organization URL to access the portal website and sign in to a portal account that has editing privileges.

    Tip: For example, the URL format would be https://gis.example.com/portal/home/.

2. On the ribbon, click the **Content** tab.

3. On the **Content** page, click the **My organization** tab.

    This displays all the content shared with the entire organization.

4. In the **Search My Organization** bar, type the name of the feature layer you published in the chapter 11 tutorial, the **Medio Public Assets web** feature layer.

    Note: Ensure that you access the feature layer and not the map image layer item. This is an important step to ensure you will be able to perform the spatial analysis.

5. Click the item. In the top-right corner of the item page, click **Open in Map Viewer**.

    Map Viewer opens with the **Medio Public Assets** web feature layer active in the map.

    To add some context to the map, you will edit the symbology of the layers.

6. In the **Layers** pane, select the **Schools** layer and on the left, click the **Styles** tab. For **Location (single symbol)**, click **Style options**. Using the **Symbol style** settings, change the point to a red **Location** pin.

7. For the **Road Centerlines** layer, change the color of the current line to dark gray.

8. For the **City Boundaries** layer, change the color of the **Solid fill** to a light blue and increase its **Fill transparency** to 25%. Change the color of the **Solid stroke** to a dark blue and increase its **Width** to **3** px.

13

9. While the **City Boundaries** layer is still selected, click the **Labels** tab on the left. Click **Add label class**. Under the **CITYDIST** label, for **Label style**, using the following settings:
   - **Font**: Arial Regular
   - **Size**: 25
   - **Font Color**: Red
   - **Halo Color**: Black
   - **Halo Size**: 2

The map should now look like this:

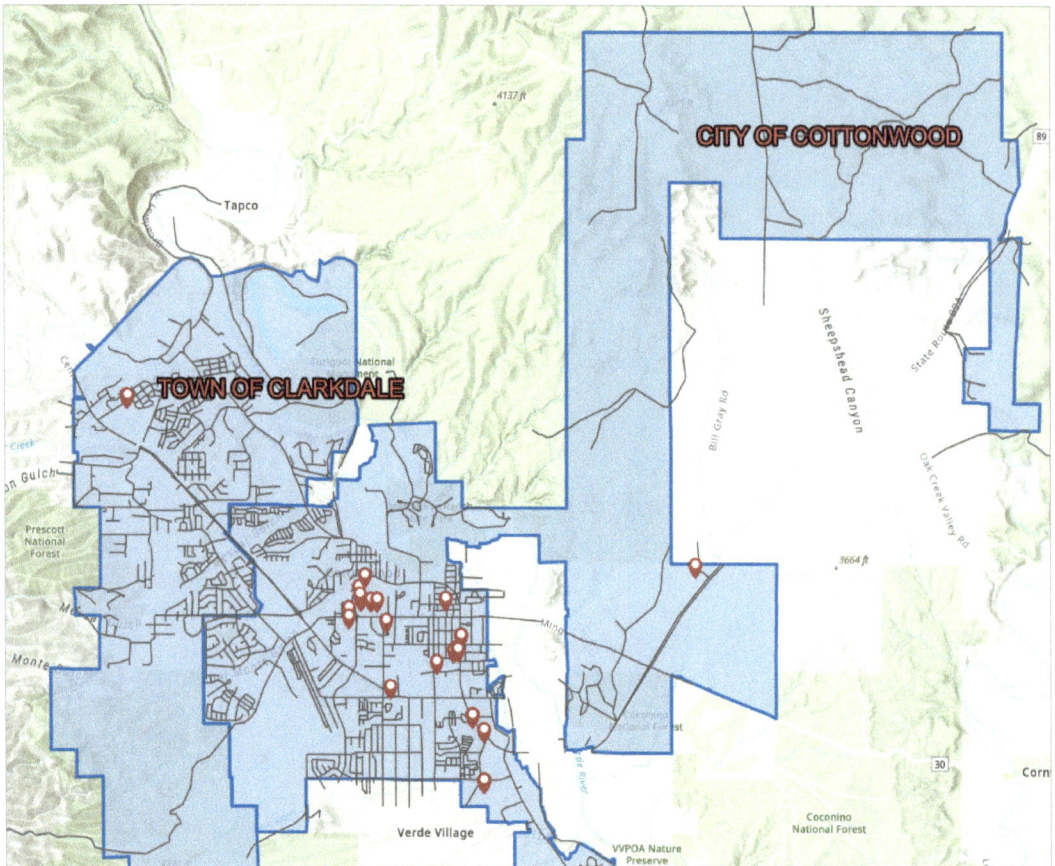

Next, you will filter the **Road Centerlines** layer to display only the road segments that are marked as unpaved.

10. In the **Layers** pane, click the **Road Centerlines** layer. Then click the **Filter** tab.

11. In the **Filter** pane, click **Add new**. Under **Condition**, apply the following expression:
- **Field name**: Street Category
- **Expression operator**: includes
- **Value**: IMPROVED UNPAVED, UNIMPROVED UNPAVED

12. When done, click **Save**.

The map now displays only the road segments with unpaved conditions.

13. In the **Layers** pane, next to the **Road Centerlines** layer, click the **Options** button (three dots) and then click **Show table**.

14. From the layer's context menu, select the **Show Table** option.

In the table, you can see the filtered road features based on the expression set earlier.

Next, you will create a half-mile buffer around each school location.

15. Click the **Analysis** tab and then click **Tools**.

16. Search for the **Create Buffers** tool and select it.

17. In the **Create Buffers** tool, apply the following settings:
- **Input layer**: Schools
- **Distance type**: Value
- **Distance values**: 0.5
- **Units**: Miles
- **Output name**: School Buffer
- **Save in folder**: County Assets

13

18. Click **Run**.

The result of the buffer analysis is a new layer that stores the 0.5-mile polygons around each school location. If you click a polygon feature on the map, you can see in the pop-up information that the polygons have inherited the school attributes.

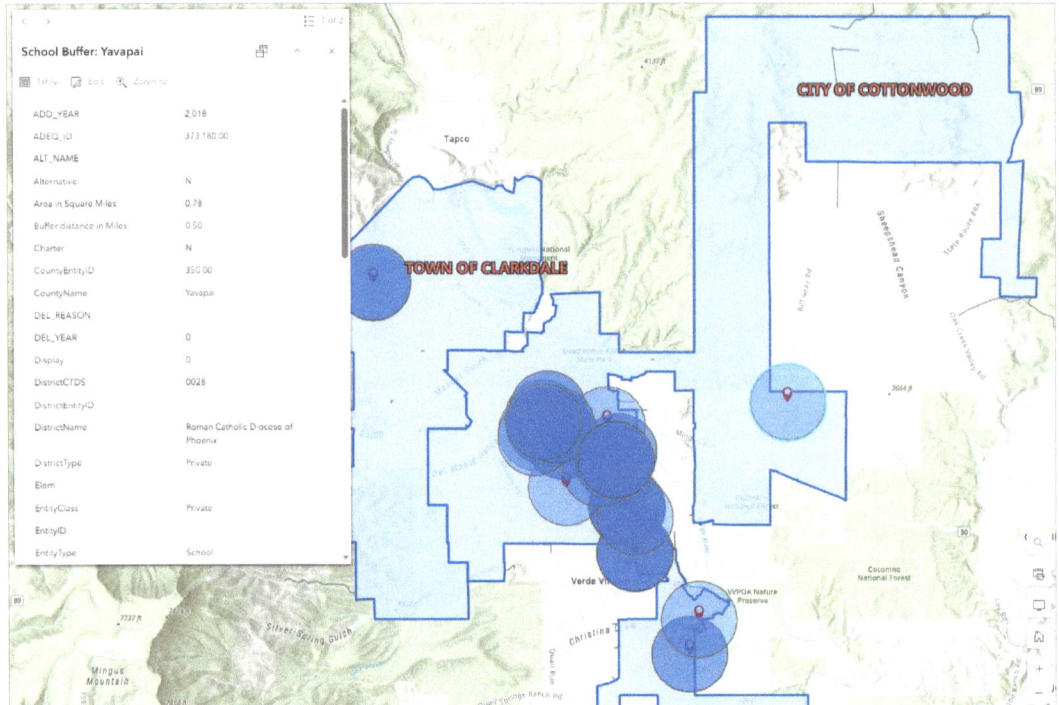

Now that the area of interest has been defined, you will identify all the unpaved roads within each buffer area. You are interested in finding not only the road location within the area but also the total number of road segments and total length of the unpaved roads in miles. To carry out these steps, you will use the **Summarize Within** tool.

19. In the **Analysis** pane, click **Tools** and search for the **Summarize Within** tool. Select the tool.

20. In the **Summarize Within** tool, apply the following settings:
- **Input features**: Road Centerlines
- **Summary polygon layer**: School Buffer
- **Output name**: SummarizeWithin
- **Save in folder**: County Assets

21. Click **Run**.

After the tool runs successfully, you can assess the newly created layer. The resulting layer is still a polygon layer with the same shape as the summary polygon layer—in this case, the **School Buffer** with additional attributes derived from calculating the statistics.

Let's review the attributes.

22. In the **Layers** pane, open the table for the **SummarizeWithin** layer.

23. In the table, scroll right toward the last two columns, which store the results of the statistics:
  - **Summarized Length in Miles** represents the total number of miles of all the unpaved roads within the buffer area.
  - **Count of Lines** represents the total number of unpaved road segments within the buffer area.

Summarized Length in Miles	Count of Lines
0.1280	2
0.0372	1
0.0000	0
0.0601	1
0.0601	1

13

To make the map a little cleaner, you will filter the resulting polygon areas based on their total length in miles. You want to display only the area with the most unpaved road segments.

24. Ensure the **SummarizeWithin** layer is selected and then open the **Filter** pane.

25. Click **Add new** and apply the following expression:
  - **Field**: Summarized Length in Miles
  - **Expression operator**: is greater than
  - **Value**: 0.7

26. Click **Save**.

The map filters, and only one polygon is now highlighted.

27. In the **Layers** pane, hover over the **School Buffer** layer and click the **Visibility** button (eye icon) to hide it.

When done, your map should have one buffer zone. This is considered the high priority area where efforts will be made to address the road conditions within proximity to the school.

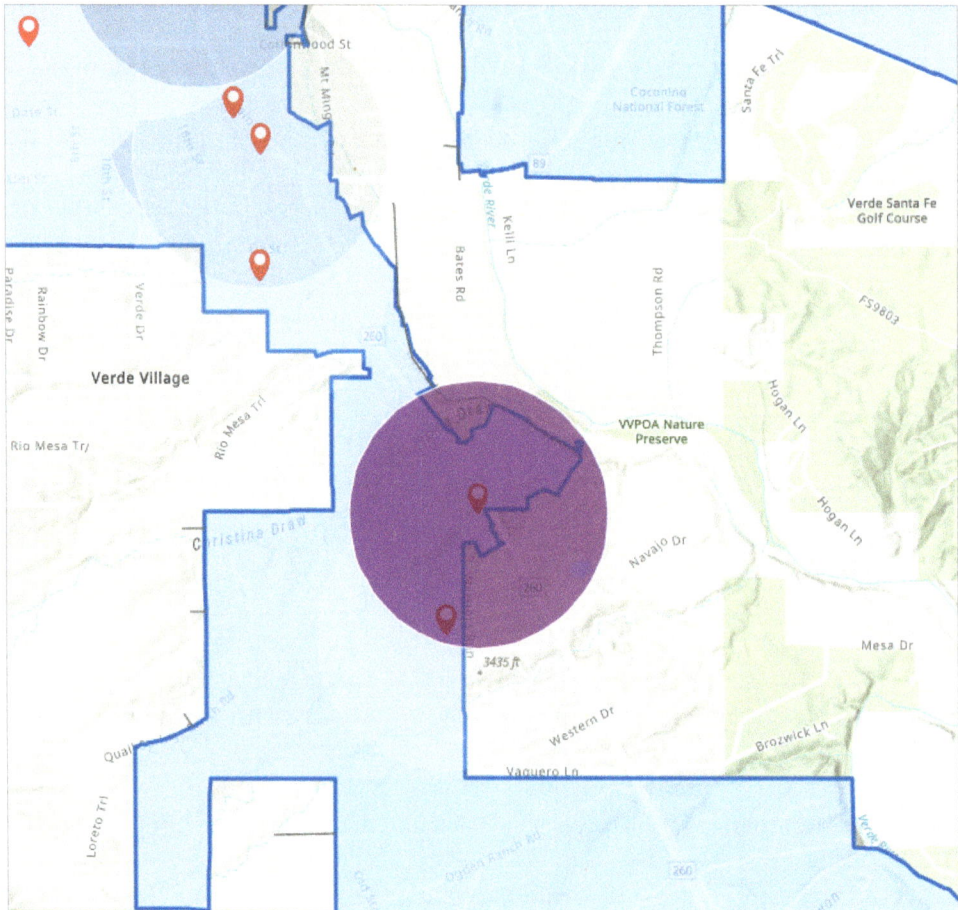

Schools with access to libraries

Next, you will assess students' access to libraries within proximity to the school. You will add a new layer published by another organization to your web map. The **Find Closest** tool will help you find the closest libraries to the school locations and assess schools without a library within two miles. Then you can identify void areas for schools without a library and also the ratio of libraries and schools.

1. In the **Layers** pane, click the **Add** button.

2. At the top of the **Add layer** pane, click **My content**, and from the drop-down menu, select **ArcGIS Online**.

    This allows you to use publicly available layers from other contributors.

3. In the search bar, type **Arizona Libraries**.

4. From the search results, locate the **Arizona Libraries** web feature layer, published by **Arizona Department of Health Services GIS**, and click **Add** to add the layer to your current map.

5. At the top of the pane, click the back arrow.

    The **Layer** pane now has **Arizona Libraries** listed as an active layer.

    Next, you will use the **Find Closest** tool to locate the libraries that are in proximity to the schools.

6. Open the **Analysis** pane. Under the **Tools** tab, expand the **Use proximity** group and open the **Find Closest** tool.

    This tool finds and ranks the features that are closest to specific locations—in this example, the school locations.

13

7. In the Find Closest tool, apply the following settings:
   - **Input layer**: Schools
   - **Near layer**: Arizona Libraries
   - **Max number of closest locations to find per unit**: 1
   - **Max search range**: 2
   - **Search range units**: Miles
   - **Output name**: SchoolsWithLibraries_2miles
   - **Save in folder**: County Assets

8. Click **Run**.

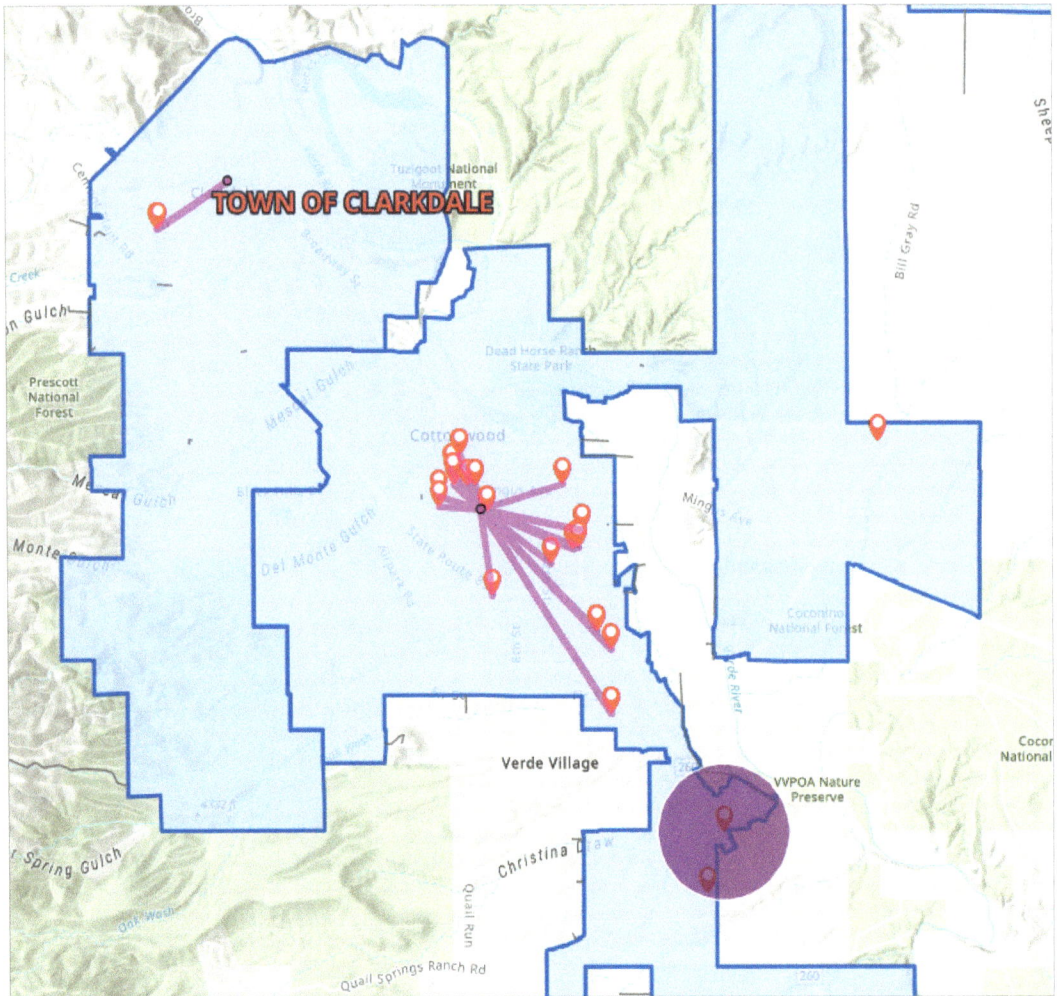

The tool generates an output with two layers that are added to the map:
- **Nearest Features** contains information about libraries that have at least one school within 2 miles radius. In this example, there are a total of three libraries.
- **Connecting Lines** contains lines connecting the input features to the closest features.

Save your findings as a web map

In this final part of the tutorial, you will prepare a web map to share with your organization. By displaying the results of your spatial analysis and adding some cartographic elements, you'll have a professional-looking map from which to base a recommendation on the school improvement project.

The **School Buffer** layer is no longer needed as the shape is preserved in the **SummizeWithin** layer.

1. In the **Layers** pane, next to the **School Buffer** layer, click the **Options** button and select **Remove**.

   Next, you will rename the **SummizeWithin** layer to something more descriptive.

2. For the **SummarizeWithin** layer, click **Options** and select **Rename**.

3. Rename the layer **High priority road area** and click **OK**.

   You will then remove the filter on the **Road Centerlines** layer and symbolize the layer based on the road category. This will help highlight the unpaved roads.

4. Ensure the **Road Centerlines** layer is selected. Open the **Filter** pane and click **Clear all** to delete the expression.

   All the road segments reappear on the map.

5. Open the **Styles** pane.

6. Under **Choose attributes**, click **Field** and select the **Street Category** field as the field you want to symbolize the layer on. Click **Add**.

7. Under **Pick a style**, click **Style options**.

13

8. Under the **Street Category** section, for each category, click the line symbol and change the **Color** using the following values:
   - **PAVED**: #757575
   - **CURRENTLY UNKNOWN**: #757575
   - **UNDEFINED**: #757575
   - **IMPROVED UNPAVED**: #e60000
   - **UNIMPROVED UNPAVED**: #e60000

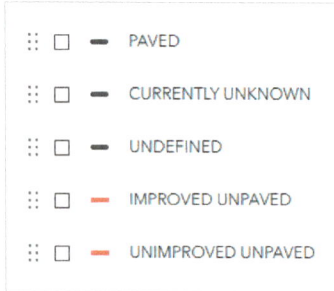

9. Click **Done**.

10. Reorder the layers in the **Layers** pane to highlight them on the map.

11. Next to each layer, use the **Reposition** button to change the order of the layers as follows:
    - **Schools**
    - **Arizona Libraries**
    - **SchoolsWithLibraries_2miles**
    - **High priority road areas**
    - **Medio Public Assets**
    - **Road Centerlines**
    - **City Boundaries**

Layers	×
:: Schools	...
:: Arizona Libraries	...
:: > SchoolsWithLibraries_2miles	...
:: High priority road areas	...
:: ∨ Medio Public Assets	...
:: Road Centerlines	...
:: City Boundaries	...

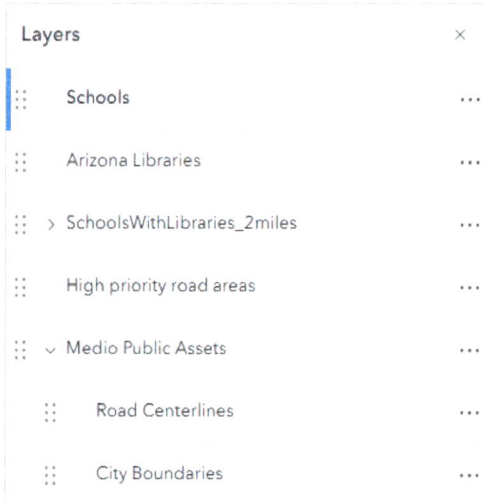

This increases the visibility of the important features on the map.

12. On the **Contents** toolbar, click **Save and open** and then click **Save as**.

13. In the **Save map** window, apply the following settings:
- **Title**: School accessibility – Spatial analysis.
- **Folder**: County Assets.
- **Tags**: schools, libraries, roads, spatial analysis, Medio county.
- **Summary**: This web map displays unpaved roads within a half-mile proximity to a school and schools without access to a library.

13

14. Click **Save**.

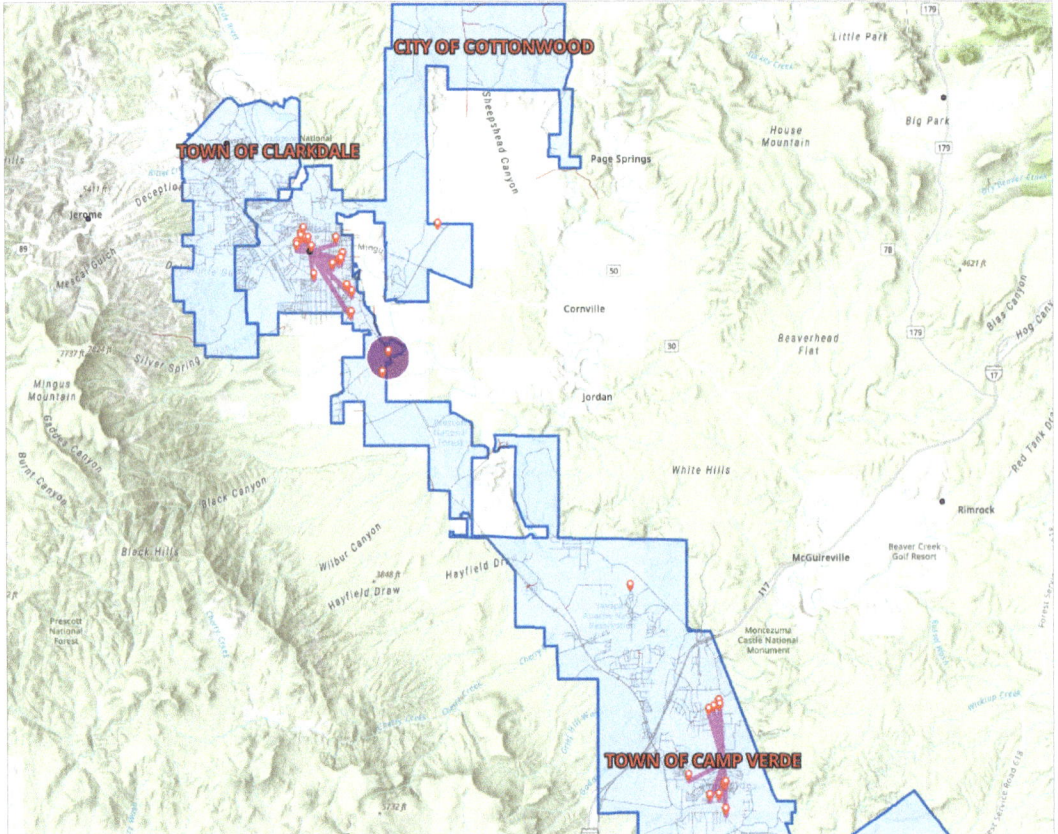

In the top-left corner of the page, the name of the map changed from Untitled to School accessibility – Spatial analysis.

Finally, you will share this map with your entire organization.

15. In the top-right corner of the page, click the three horizontal lines and click the **Content** tab.

On the **My Content** tab, the School accessibility – Spatial analysis map is listed as a web map item.

16. Click the title to open the item page.

17. Confirm the item summary is the same text you added when the web map was saved.

18. On the left side of the page, scroll down on the page until you see the **Share** option and click the **Edit** button (pencil icon).

19. In the **Share** window, change the sharing level to **Organization** and then click **Save**.

This ensures all members of your organization will have access to this web map.

A warning message appears on the screen informing you that the web map references other items that might not be visible to your organization members because these items have a different level of sharing permissions.

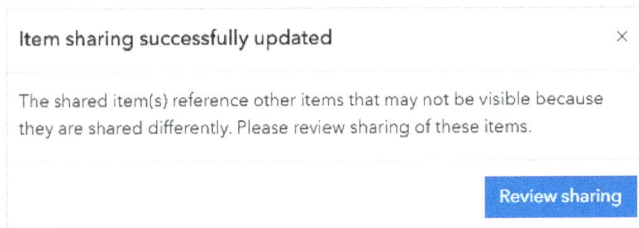

Item sharing successfully updated	✕
The shared item(s) reference other items that may not be visible because they are shared differently. Please review sharing of these items.	
	Review sharing

20. Click **Review sharing** to inspect the layers.

It looks like the layers generated from your spatial analysis results are still shared only with the owner of the item. This means that although the web map is shared with the organization, the resulting layers will not be visible in the map.

13

21. Click **Update sharing** to synchronize the sharing level of these layers to match the web map.

The web map together with the referenced layer and the hosted feature layers are now shared and available to the entire organization.

Take the next step

If you have access to another user in your ArcGIS Enterprise portal, use the other user to test the accessibility of the web map from the **My Organization** page.

Summary

In this chapter, you delved into the capabilities and benefits of spatial analysis within ArcGIS Enterprise. You learned how to create a new web map and conducted spatial analysis to assess school accessibility. By using the Summarize Nearby tool, you identified areas with the most unpaved roads, calculating the total length of unpaved roads within a half-mile buffer around schools. You then enhanced your analysis by adding the Libraries hosted feature layer from ArcGIS Online. Using the Find Closest tool, you pinpointed schools that lack nearby libraries. The results of your analysis were saved into a new web map, which you shared with your organization for easy access and collaboration.

In the next chapter, you will present your insights to a larger group of stakeholders within the organization, focusing on county development and resource allocation. Because not all stakeholders are GIS professionals, you will display your analysis results using ArcGIS Dashboards, creating interactive and user-friendly tools to facilitate understanding and decision-making.

CHAPTER 14

Communicating your analysis results in a dashboard

Objectives

- Create a dashboard from scratch.
- Configure dashboard information elements.
- Configure dashboard selectors.
- Configure map actions.

Introduction

Organizations increasingly rely on data to solve problems, identify trends, communicate results, collaborate on projects, and make informed decisions. Optimizing data presentation enhances the experience for your target audience and facilitates effective information sharing. In the GIS world, integrating data and maps plays a crucial role in decision-making. The ArcGIS system offers a wide range of app builders that improve how organizations manage and present spatial data. These tools enable users to create interactive web apps without extensive coding knowledge, using a no-code and low-code approach that saves time and resources.

Esri offers four app builders, each supporting specific business needs:

- **ArcGIS Instant Apps** allows users to quickly create web apps that facilitate interaction with maps and data. With various templates and configuration options, users can build apps tailored to their needs without any coding. Instant Apps supports accessibility across devices and provides a user-friendly interface for creating and sharing apps within minutes.
- **ArcGIS Dashboards** enables users to create interactive and informative data dashboards that present geographic information and data on a single screen. Dashboards

display multiple visualizations, such as maps, charts, and gauges, providing a comprehensive view of the data. This app builder is essential for monitoring events, making decisions, and informing others in real time.

- **ArcGIS StoryMaps** allows users to transform their maps and GIS work into engaging narratives. By combining maps, text, images, videos, and other multimedia content, users can create interactive stories that inform, inspire, and engage their audience. ArcGIS StoryMaps is ideal for sharing complex information in a compelling and accessible format.
- **ArcGIS Experience Builder** provides a flexible platform for creating web apps that integrate 2D and 3D data. Using Experience Builder's drag-and-drop interface, users can design custom templates and build mobile-adaptive apps without writing code. Experience Builder supports a wide range of layouts, content, and widgets, allowing users to create unique web experiences tailored to their specific needs.

All these app builders support seamless integration with ArcGIS Online and ArcGIS Enterprise, facilitating the sharing of maps and data across platforms. This capability allows organizations to use GIS data in a more interactive and impactful way, driving informed decision-making and collaboration.

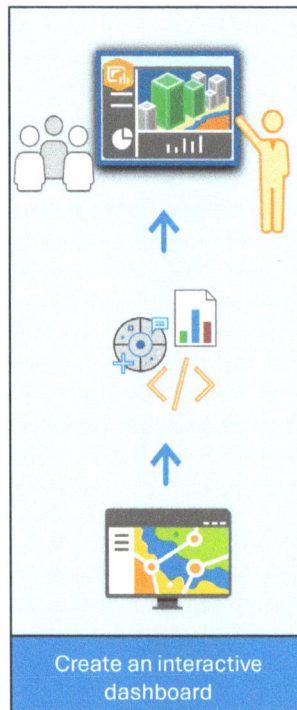

Figure 14.1. Sharing data through a dashboard.

Tutorial 14: Make a dashboard for public presentation

In this tutorial, you will prepare for a public presentation to showcase the results of your spatial analysis. You will advocate for funds to address road conditions and library issues in Medio County to improve school accessibility. The audience for this presentation includes individuals who are not GIS specialists, so you will need to convey the results of your spatial analysis in a more interactive and easy-to-read format. To achieve this goal, you will use Dashboards, an easy-to-use app builder, to create a dashboard that presents technical information in an accessible way. You will learn how to incorporate dashboard elements, such as titles, indicators, detailed lists, charts, and actions, to enable users to interact with the data effectively.

Configure a new dashboard

You will begin with an empty dashboard template, adding a title and selecting a thematic color. Because the map is the central element of the dashboard, you will define various actions that users can perform on the map to enhance interactivity and engagement.

1. Ensure you are connected to your organization's ArcGIS Enterprise portal.

   > *Tip: For example, the URL format would be https://gis.example.com/portal/home/.*

2. Sign in as a portal user that has at least editing credentials.

   This ensures that you can edit the web feature layer.

3. In the top-right corner of the page, click the **App Launcher** button (nine dots).

4. From the list of apps, select **Dashboards** to open the app on a new tab.

5. Click **Create dashboard**.

6. In the **Create new dashboard** window, type the following information:
   - **Title**: Medio County Public Assets Report.
   - **Tags**: roads, libraries, schools, annual report, Medio County.
   - **Summary**: This dashboard is used to support the school accessibility project in Medio County.
   - **Folder**: County Assets.

14

7. Click **Create dashboard**.

The dashboard appears with an empty template. You will next use the toolbar on the left to customize the dashboard.

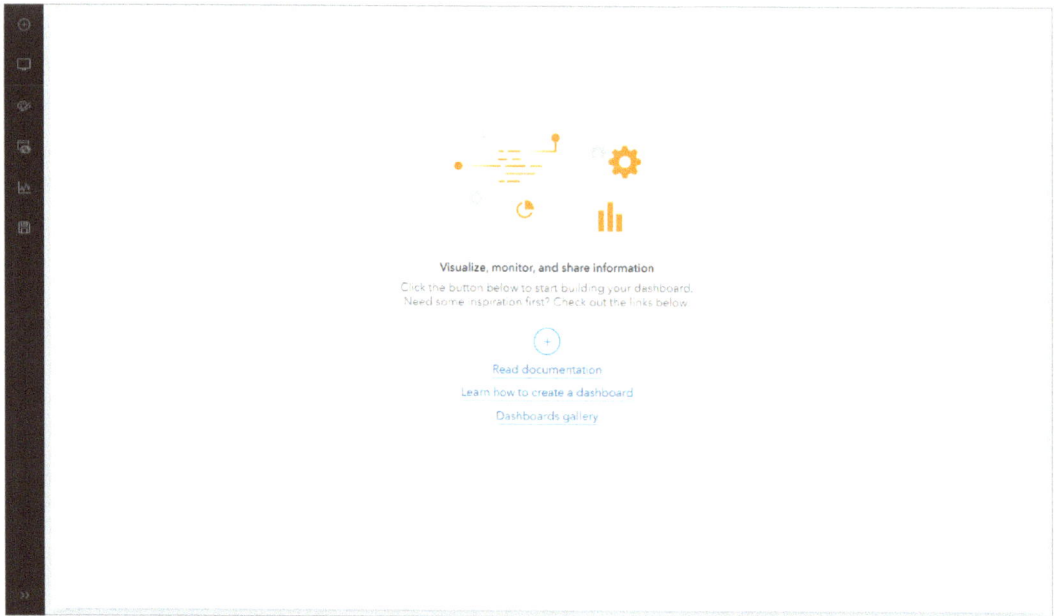

> *Tip: At the bottom of the dashboard toolbar, you can click Expand or Collapse arrows to show or hide the tool names on the toolbar.*

8. On the dashboard toolbar, click the **View** tab. In the **View** pane, click the **Header** tab and click the **Add header** button.

The header acts as a title for the dashboard.

9. Apply the following settings for the header:
   - **Title**: Medio County Annual Report
   - **Text color**: #ffffff
   - **Foreground color**: #42888B

10. Click **Done**.

11. On the dashboard toolbar, click the **Theme** tab. Change the theme to **Dark** and click **Customize selected theme**.

12. Under **Colors**, click **Customize**.

13. Expand the **Colors** group. Change the **Text color** to #ffffff.

14. Expand the **Advanced colors** group. Change the **Background color** to #ababab.

15. From the toolbar, click **Save** to commit the changes you made.

Now that you have configured the overall appearance of the dashboard, you will add the **School accessibility – spatial analysis** web map and start populating the dashboard with information from the map.

Add map element

You will first add the spatial analysis web map and then configure the dashboard to display the total number of schools in the county with access to a library, as well as the total number of libraries. Next, you will include a histogram to illustrate the current road conditions and an interactive list of unpaved roads.

1. On the dashboard toolbar, click the **Add element** tab.

2. In the middle of the dashboard, click the **Add** button (plus sign icon) located in the middle of the dashboard.

3. From the list of elements, select **Map**.

4. In the **Select a map** window, navigate to the **My content** or **Shared content** tab. Then select the **School accessibility – spatial analysis** web map.

You can configure how the map will appear on the dashboard.

14

5. For **Settings**, change the **Scalebar** to **Line** and then turn on the following map tools:
   - **Search**
   - **Legend**
   - **Basemap switcher**
   - **Zoom in/out**

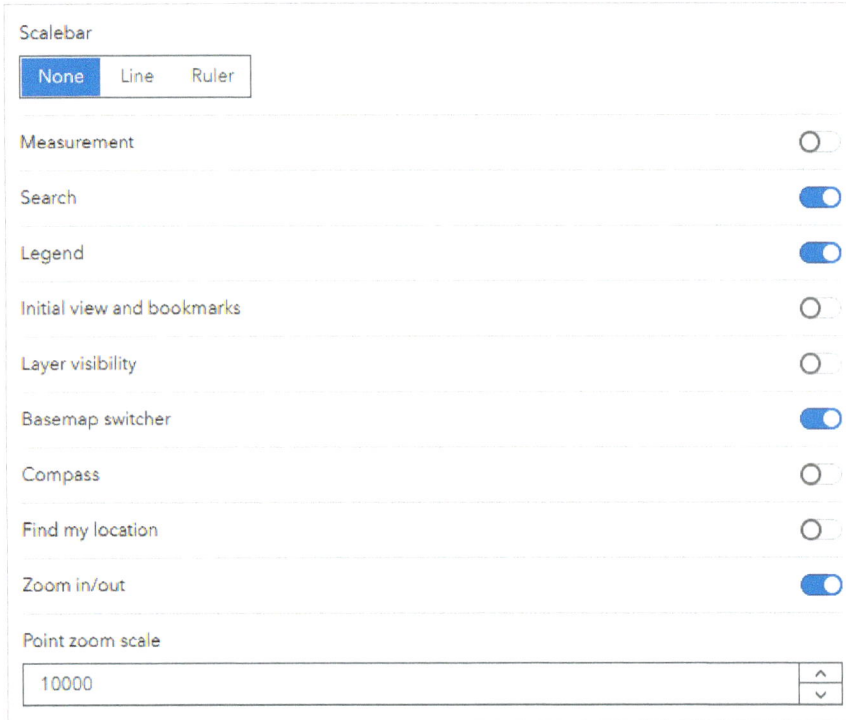

Scalebar

| None | Line | Ruler |

Measurement ⚪

Search 🔵

Legend 🔵

Initial view and bookmarks ⚪

Layer visibility ⚪

Basemap switcher 🔵

Compass ⚪

Find my location ⚪

Zoom in/out 🔵

Point zoom scale

10000

6. When finished, click **Done**.

   The web map appears in the center of the dashboard. You will also notice that the interactive tools that you enabled earlier have appeared in the top-right corner of the map. These options allow your audience to interact with the map.

7. In the top-right corner of the map, click the **More tools** button.

8. Click the **Basemap** tool and select **Light Gray Canvas**.

   Next, you will add a chart to display the road conditions.

Add Serial chart element

1. Click the **Add element** tab. On the right side of the dashboard, click the **Add** button and then select the **Serial chart** element.

2. In the **Select a layer** window, expand **Medio Public Assets** and select the **Road Centerlines** layer.

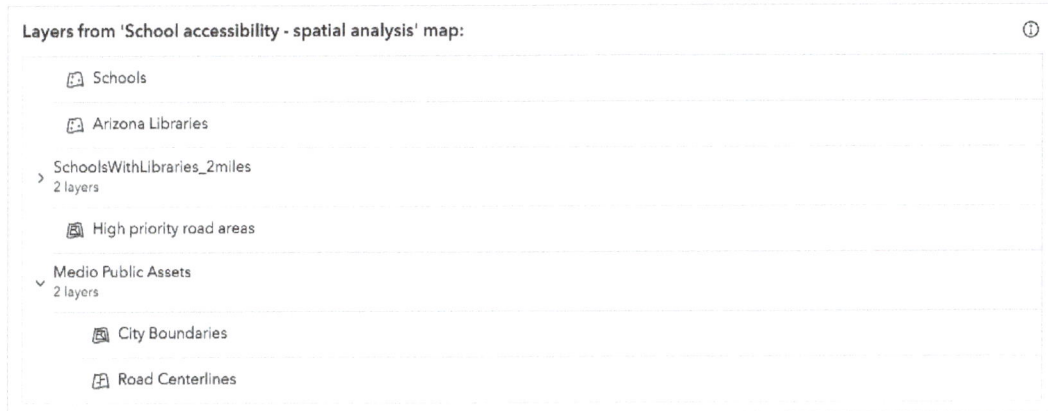

Layers from 'School accessibility - spatial analysis' map: ⓘ

    ▢ Schools

    ▢ Arizona Libraries

> SchoolsWithLibraries_2miles
> 2 layers

    ▣ High priority road areas

∨ Medio Public Assets
  2 layers

    ▣ City Boundaries

    ▢ Road Centerlines

3. In the **Serial chart** window, apply the following **Data options** settings:
   - **Categories from**: Grouped values
   - **Category field**: Street Category
   - **Statistics**: Count

4. Next to **Filter**, click the Filter button. Create the following expression:
   - **Field**: Street Category
   - **Operator**: include
   - **Value**: IMPROVED PAVED, UNIMPROVED PAVED, PAVED, CURRENTLY UNKNOWN

5. Click **Done**.

6. On the left of the window, click the **Chart** tab. Change the **Orientation** to **Horizontal**.

7. Click the **Category axis** tab and change the **Title** to Current road conditions.

8. Scroll down and expand the **Labels** group. Change the **Placement** to **Wrapped** and then click **Load categories**.

14

9. Rewrite the labels as follows:
   - **CURRENTLY UNKKNOWN**: Unknown
   - **PAVED**: Paved
   - **IMPROVED UNPAVED**: Unpaved improved
   - **UNIMPROVED UNPAVED**: Unpaved

10. Click the **Value axis** tab. Change the **Title** to Total number of road segments and then turn on the **Logarithmic** option.

11. Click the **Series** tab and change the **Color** to #42888B.

12. Click the **General** tab. Scroll down and turn on the **Last update text** option.

    This informs your audience when the chart was last updated.

13. Once you have finished with the chart settings, click **Done**.

    The chart should have four horizontal bars representing the current road conditions for the assigned street categories.

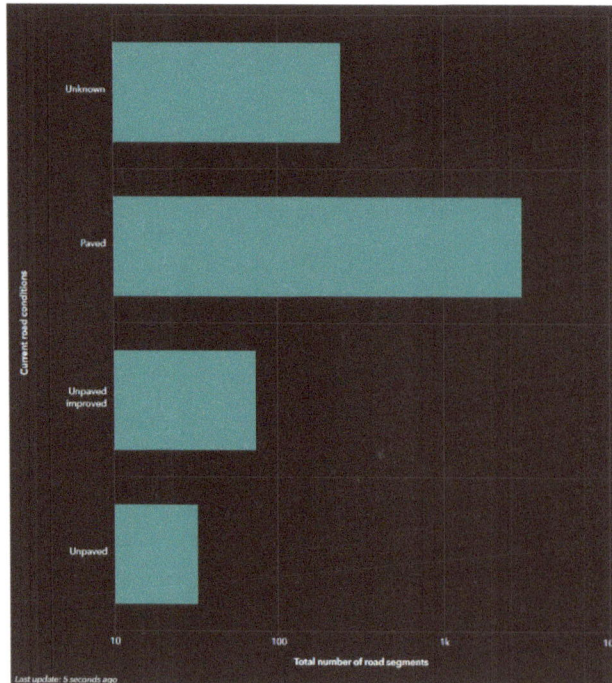

> *Tip: To change an existing element in the dashboard, hover over the chart element's top-left corner where the Options button appears. You can drag the element to a different location, change the con-figuration settings, duplicate the element, or delete the element.*

Next, you will resize the chart so that the map is the focus of the dashboard.

14. Hover over the vertical divider between the map and the chart until the pointer changes to cross hairs. Drag the divider toward the right until the percentage reads **68.0%**.

Add Indicator element

Next, you will configure two indicators to show the total number of schools with library access and total number of libraries within Medio County.

1. Click the **Add element** tab and add an element below the chart. Select the **Indicator** element.

2. From the **Select a layer** window, expand the **SchoolwithLibraries_2miles** layer and select the **Connecting Lines** layer.

    This ensures you are displaying the total number of schools that have a library nearby.

3. For **Data options**, confirm the **Statistic** parameter is set to **Count** and the **Field** is the **objectID**.

4. Click the **Indicator** tab. For **Middle text**, change both the **Text color** and the **Outline** color to #2b5d5e.

5. For **Icon**, click **Add icon**. Expand the **Points of interest** group and select the graduation cap icon. Click **OK**. Change the **Fill** icon color to #FFAA00.

14

6. Click the **General** tab. Change the **Title** to Schools with libraries within 2 miles.

7. Change the **Foreground color** to #42888B.

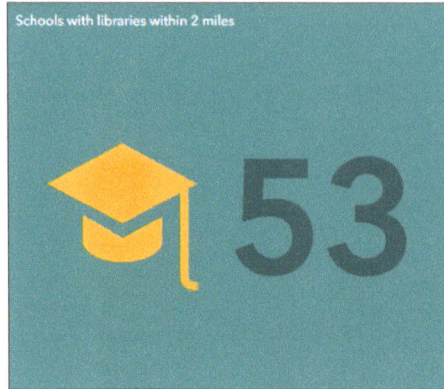

Next, you will add another **Indicator** element to show the total number of libraries in Medio County. But instead of creating it from scratch, you will duplicate the first indicator.

8. Hover over the top-left corner of the indicator and click **Duplicate**.

   An identical indicator appears under the school one.

9. Hover over the duplicated indicator and click **Configure**.

10. For **Data options**, next to **Layer**, click **Change** and select the **Nearest Features** layer.

    This layer shows the libraries that are near the schools within the Medio County boundary.

11. Click the **Indicator** tab. Next to **Icon**, click **Change**. From the **Points of interest** group, select the two-story building with a flag on top. Click **OK**.

12. Click the **General** tab and edit the **Title** to Libraries in Medio County. Click **Done**.

13. Hover over the new indicator. Click **Drag item** to drag the indicator to the right of the previous indicator.

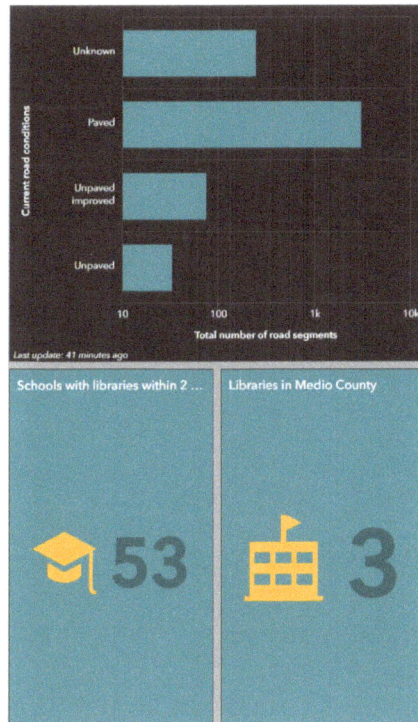

Add List element

The final element of the dashboard is a list with the names of all the unpaved roads. The list will be interactive, allowing users to click an item in the list. When users do this, the map will zoom to that specific road feature.

1. Add the **List** element to the left side of the map. Select **Roads Centerlines** as the source layer.

2. Add the following filter to display only unpaved roads: **Street Category include IMPROVED UNPAVED, UNIMPROVED UNPAVED.**

Filter

| Street Category | abc ⌄ | 🗑 |

| include | ⌄ |

| IMPROVED UNPAVED ✕   UNIMPROVED UNPAVED ✕ | ⌄ |

AND   OR

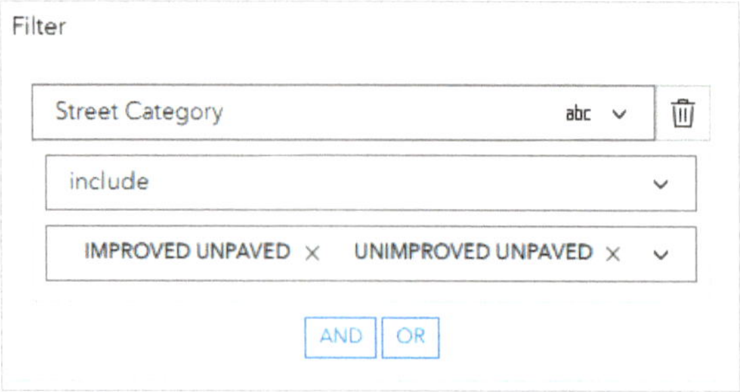

3. On the **List** tab, for **Template**, type **Stree**t at the end of the current text.

🔗 Source

{field/street_name} Street

This adds the word Street after all the street names in the list.

4. Scroll down and change the **Icon** to **None**.

5. On the **General** tab, change the **Title** to List of unpaved roads.

To make the dashboard element interactive, you will set up an action that links the element of the dashboard with the map,

6. Click the **Actions** tab.

7. Expand **Flash**, **Show pop-up**, and **Zoom**. For each section, turn on the School accessibility – spatial analysis web map.

8. Click **Done**.

9. To test the action, click on one of the street items from the list.

The map zooms in to that street feature, and the pop-up is active so that your audience can inspect the feature attributes.

10. As final touch-ups, adjust the vertical and horizontal borders of each element so that the map is the focus of the dashboard.

11. When finished, on the dashboard toolbar, click **Save**.

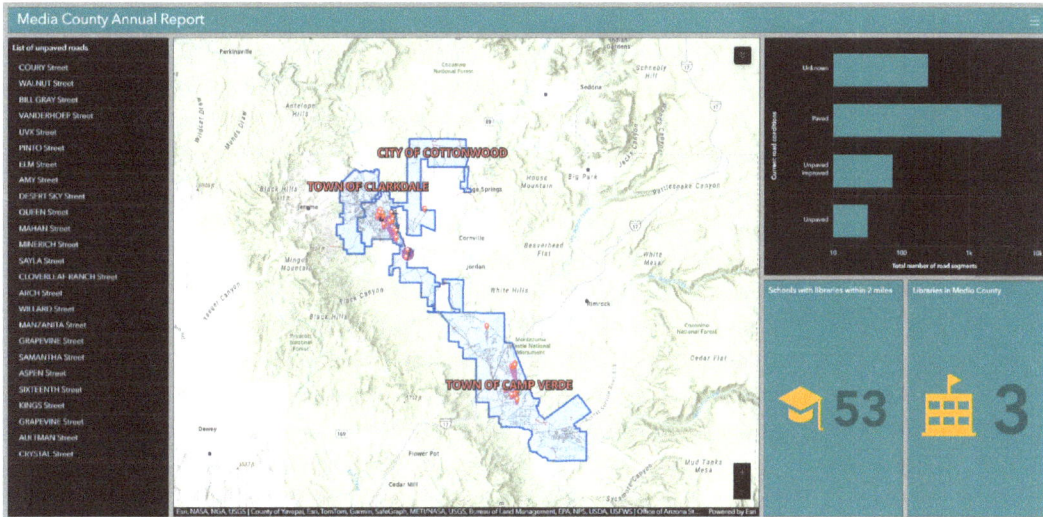

Share your dashboard with the organization

The final step of this workflow is to set your dashboard share settings so that anyone in your organization can view the dashboard. You will also test the URL to be sure the app works as expected for the annual report meeting.

1. At the top-left-corner screen, click the horizontal three lines and select **Dashboard item details**.

The item page for your dashboard appears.

2. Click the **Share** button.

3. In the **Share** window, select **Organization** and click **Save**.

4. Click **Open Dashboard**.

14

Your dashboard appears on a new tab. This is what the members of your organization will see when you share the link with them.

5. Explore and test the dashboard to ensure it is functioning as you expect.

If your dashboard appears and functions as expected, it is ready for the annual presentation meeting.

Summary

In this chapter, we explored the needed steps to effectively communicate analysis results through the creation of a dashboard. By following the objectives outlined, you learned how to create a dashboard from a template, configure various dashboard information elements to highlight details from your spatial analysis, and implement map actions to make the entire experience more engaging for your audience.

In the fifth and final tutorial of part 3, you will collaborate with a local contractor firm to start the construction phase for some of the unpaved roads around school locations. The local firm hosts its data in ArcGIS Online, where it can better serve the size and purpose of the firm's work. To share the Medio Public Assets web feature layer from your department's ArcGIS Enterprise to an ArcGIS Online organization, you will establish a distributed collaboration.

CHAPTER 15

Establishing a distributed collaboration

Objectives

- Create and join a distributed collaboration.
- Understand the role of the host, guest, and workspace.
- Synchronize edits between organizations.
- Leave and end a collaboration.

Introduction

Distributed collaboration is a technology that allows multiple organizations, often located in different geographic regions, to work together on a common project or goal. The collaboration can be established between ArcGIS Enterprise organizations or an ArcGIS Online and ArcGIS Enterprise organization. This allows for content, such as items, feature layers, web maps, web apps, and more, to be shared between these organizations. This approach uses cloud-based services and web GIS to ensure that all collaborators have access to the most up-to-date information, supporting a more integrated and efficient workflow.

This approach significantly enhances the ability to make informed decisions quickly and accurately. By breaking down silos and enabling real-time data sharing, organizations can respond more effectively to emergencies, plan urban development more efficiently, and manage natural resources more sustainably. Additionally, distributed collaboration promotes transparency and accountability, because all stakeholders have access to the same data and insights, reducing the risk of miscommunication and errors. In essence, it allows organizations to harness the full potential of their GIS data, driving innovation and improving outcomes across sectors.

Figure 15.1. Distributed collaboration diagram.

Create a new collaboration

When you plan a collaboration between organizations, it is important to understand the data considerations and concepts of collaboration, such as the roles of the host organization, guest organization, and workspace.

Before creating the collaboration for the web feature layer you published in the previous chapter, it is necessary to confirm that the web feature layer and its three feature classes have the right capabilities enabled to support data sharing in a distributed collaboration. When you published the Medio Public Assets web feature layer to ArcGIS Enterprise, you enabled the following capabilities:

- **Archiving**: To support syncing operations and distributed collaboration
- **Replica tracking**: To support bidirectional syncing in a distributed collaboration
- **Global IDs**: Enabled as part of the publishing process
- **Sync capability**: Enabled as part of the web feature layer configuration

These are the minimum requirements to share your data with another organization using distributed collaboration.

Next, you will review the responsibilities of the host and guest organizations in a collaboration.

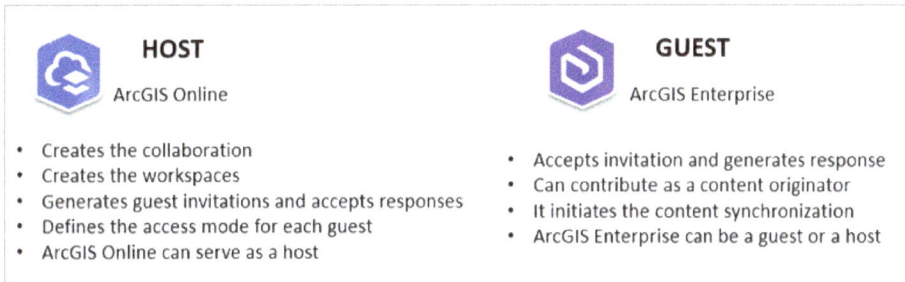

HOST
ArcGIS Online

- Creates the collaboration
- Creates the workspaces
- Generates guest invitations and accepts responses
- Defines the access mode for each guest
- ArcGIS Online can serve as a host

GUEST
ArcGIS Enterprise

- Accepts invitation and generates response
- Can contribute as a content originator
- It initiates the content synchronization
- ArcGIS Enterprise can be a guest or a host

Next, you will review the workspace properties in this chapter's tutorial.

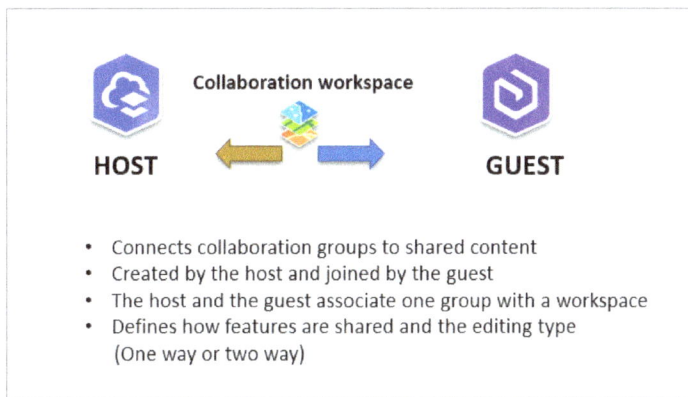

Collaboration workspace

HOST

GUEST

- Connects collaboration groups to shared content
- Created by the host and joined by the guest
- The host and the guest associate one group with a workspace
- Defines how features are shared and the editing type
  (One way or two way)

Tutorial 15: Share data through a distributed collaboration

15

In this final tutorial of part 3, you will collaborate with a local road company to begin work on some of the unpaved roads. The dashboard created in the previous tutorial was a success, and the stakeholders have approved funds for roadwork. The local contractor firm undertaking this project has an ArcGIS Online account to host all its data.

Now you will establish a distributed collaboration between your ArcGIS Enterprise organization and Medio Contractors LLC's ArcGIS Online account. The first synchronization will make the **Medio Public Assets** layer available to the ArcGIS Online organization. The organization will use this layer to make edits as the project progresses. During the scheduled workspace synchronizations, the data collected becomes available to the ArcGIS Enterprise organization. This process repeats throughout the duration of the project. Once the project is complete, the guest leaves the collaboration, and the host can delete it.

You will start with the host, the ArcGIS Online organization. The host creates the distributed collaboration, the workspace, and invites the guest organization to join the collaboration. For this tutorial, you will need an ArcGIS Online account with **Administrator** privileges.

1. In a new browser, navigate to **https://www.arcgis.com**.

2. Sign in as an ArcGIS Online member who has administrative privileges to manage the organization's collaborations.

3. Once connected, on the navigation bar, click the **Organization** tab and go to **Settings**.

4. On the left, click the **Collaborations** tab.

   You will see two types of collaborations that can be created: (1) a partnered collaboration, which is established between two ArcGIS Online organizations, and (2) a distributed collaboration, which is established between an ArcGIS Online and an ArcGIS Enterprise organization to share content.

5. Scroll down to the **Distributed** section. Click **Create a distributed collaboration**.

6. In the **Create a distributed collaboration window**, provide the following name and description for the collaboration:
   - **Collaboration Name**: Roadwork project.
   - **Collaboration Description**: This is a distributed collaboration between Medio County and Medio Contractors LLC to support roadwork around school locations.

7. Click **Next**.

8. Provide the following name and description for your collaboration's initial workspace:
   - **Workspace Name**: Medio roadwork.
   - **Workspace Description**: This workspace is to share the Medio County data and give access to the ArcGIS Online organization to make edits to the legacy data as the roadwork advances.

9. Click **Next**.

Next, you will create a new group that will be associated with the collaboration workspace. Once the ArcGIS Enterprise workspace shares the **Medio Public Assets** data, it will be stored into this group.

10. For **New Group** name, type **Medio County Group** and click **Next**.

In this step, you will determine how the web feature layers will be shared between these two organizations. There are two options:
- **Share as a copy**: In this case, the data is copied from the source organization to the recipient organization as a hosted feature layer. The web layer needs to support syncing, because the edits will be synced on a set schedule.
- **Share as a reference**: Here, the layer is created in the recipient organization that references back to the source layer. Edits are available immediately.

11. Select the **Copies** option. Check the box for **Allow two-way sharing of feature service edits to eligible participants**.

This ensures that both organizations can make edits.

Feature layers and views in my portal are sent as
- ◯ References
- ⦿ Copies

Collaboration participants receive feature layers and views as hosted feature layers that are periodically synced. Ensure that you check 'Enable Sync' on the source feature layer's or view's item settings to send it as a copy. Learn more
- ☑ Allow two-way sharing of feature service edits to eligible participants
- ☑ If unable to share as copies share as references

12. Click **Save and Invite**.

13. For the **Guest Organization URL**, add the ArcGIS organization URL you used for publishing the Medio County data.

14. For **Guest Organization Access to Workspaces**, click the drop-down menu and select **Send and Receive Content**.

15. Click **Send Invitation**.

15

An .invite file is generated in the **Downloads** folder. You will use this file as the guest organization to accept the invitation to join the collaboration.

Join a collaboration

In this section, you will act as a member of the ArcGIS Enterprise organization—the guest. You will create a new group and the data of interest to that group. Then, you will accept the host's invitation to join the collaboration, join the workspace, and determine the sync interval.

1. In a new browser, connect to your ArcGIS Enterprise portal and sign in with an administrator account.

2. On the navigation bar, click the **Groups** tab and then click **Create group**.

3. In the **Create a group** window, apply the following settings:
   - **Name**: Collaboration group.
   - **Summary**: This group is used in the Roadwork collaboration.

4. Leave the remaining parameters as default and click **Save**.

5. On the **Collaboration group** item page, click **Add items to group**.

6. Add the **Medio Public Assets** web feature layer.

   This will be the layer used to make edits.

   Now that the group is created and the layer is prepared, you will accept the collaboration invitation.

7. Click the **Organization** tab and then go to **Settings**.

8. On the left, click the **Collaborations** tab. Under **Distributed**, click **Accept Invitation**.

9. In the **Accept Collaboration Invitation** window, click **Choose File** and navigate to the .invite file generated when the collaboration was created. Click **Open**.

10. When done, click **Accept Invitation**.

11. In the **Save Invitation Response** window, click **Save Response**.

This downloads a file that contains your response to the invitation from the ArcGIS Online organization.

After accepting the invitation, the **Roadwork projec**t appears as an active collaboration on the guest side.

Collaboration Name	▲ Participation	Creation Date	Action
Roadwork project	Guest	2025	⚙ ▾

Next, you will join the collaboration workspace.

12. Click on the link for the **Roadwork project** collaboration.

This takes you to the list of workspaces.

13. Click the **Action** icon and then click **Join workspace**.

On the guest side, the workspace will be linked to an existing group, the **Collaboration group** you created earlier.

14. Select **Existing Group** and change it to the **Collaboration group**.

15. Change the **Feature layers and views** option to be sent as **Copies**.

16. Check the box for **Allow two-way sharing of feature service edits to eligible participants**.

The final step as a guest is to define the feature layer sync interval.

17. Scroll down. For **Start syncing at**, type **05:00**, and for **repeat every**, select **24 hours**.

18. Click **Join Workspace**.

To finalize the process, the host, the ArcGIS Online organization, receives the response file from the guest and accepts it.

19. Return to the ArcGIS Online browser and navigate to the **Collaborations** page.

15

20. Under the **Distributed** collaborations, click the link for **Roadwork project** collaboration.

21. Click **View Guests**.

   The ArcGIS Enterprise organization is listed as a guest, but the collaboration invitation is still pending.

22. Click the **Action** icon and click **Accept Guest Organization**.

23. Click **Choose File** and navigate to the **.response** file received by the guest. Click **Open**.

24. Click **Accept Guest Organization**.

   The guest organization status is set to **Active**. The collaboration has successfully started, and the organizations are ready to share data.

Make edits and sync workspace

In this section, the host organization will edit the **Road Centerlines** layer. These edits will become available to the guest organization at the next scheduled sync. When the guest organization joined the workspace, a sync process was automatically initiated, making the feature layer available to the host organization, ArcGIS Online. You will confirm this by checking the **Medio County Group** from the host organization.

1. In the ArcGIS Online browser, on the navigation bar, go to **Groups**. Then locate the **Medio County Group**.

2. On the **Groups** page, confirm that **Medio Public Assets** web feature layer is now listed under **Recently added content**.

   Next, you will use this layer to make edits to unpaved roads within school proximity.

3. Click **Medio Public Assets** and then click **Open in Map Viewer**.

4. In the **Layers** pane, expand **Medio Public Assets** and click the **Road Centerlines** layer.

   This activates the layer for editing.

5. On the **Analysis** toolbar, click the **Edit** tab.

6. Because you are not creating new features but editing existing ones, in the **Editor** pane, click the **Select** icon.

7. Select any unpaved road within school proximity and change the **Street Category** field value to PAVED. Click **Update**.

For example, the UVX Road is in close proximity to the Cornerstone Christian Academy (red pin) and is unpaved.

8. When finished, in the **Layers** pane, next to **Road Centerlines**, click the **Options** button (three dots). Click **Save**.

This ensures the edits made are committed.

> *Tip: The web layers shared through a distributed collaboration can also support field data collection apps, such as ArcGIS Field Maps. Because the web layers are sync enabled, editors can collect data on mobile devices and sync it with ArcGIS Online.*

Next, you will manually synchronize the workspace as the guest organization, ArcGIS Enterprise, and confirm that the changes made by the host organization are now available.

9. Activate the **ArcGIS Enterprise** browser and go to **Collaborations**.

10. Click the **Roadwork project**, which will take you to the list of active workspaces.

11. Go to the **Action** icon and click **Sync workspace**.

12. Click the **Medio roadwork** workspace to see the sync status.

The current status is **In Progress**. Once the status is changed to **Succeeded**, you can confirm the edits are now available to the guest organization, ArcGIS Enterprise.

Once edits are synchronized with ArcGIS Enterprise, the changes are reflected not only in the **Medio Public Assets** source web feature layer but also in all applications powered by this web feature layer. This means that the edits made to the road segments will be immediately visible in the **School Accessibility – Spatial Analysis** web map.

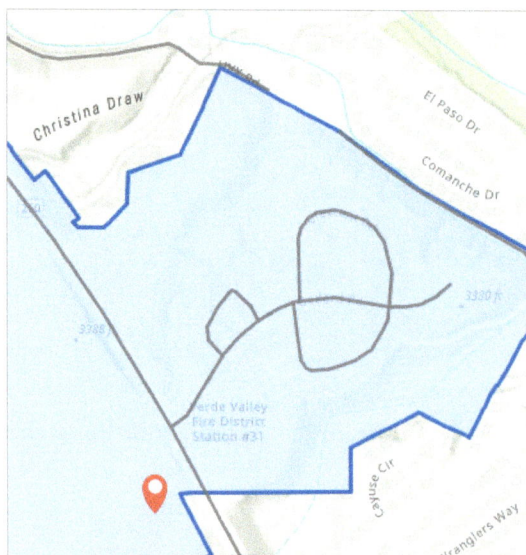

The changes are also available in the **Medio County Annual Report** dashboard. If you check the list of unpaved roads, you will confirm that these two road segments are no longer listed. This confirms that the edits made by the host organization are also reflected in the dashboard.

List of unpaved roads	List of unpaved roads
COURY Street	COURY Street
WALNUT Street	WALNUT Street
BILL GRAY Street	BILL GRAY Street
VANDERHOEF Street	VANDERHOEF Street
UVX Street	PINTO Street
PINTO Street	

Additionally, because the data was published by referencing a registered data store, these edits are also updated in the enterprise geodatabase. Once you connect to your ArcGIS Pro project and access the Medio County data from the database connection, you will confirm that the changes are reflected to the source enterprise geodatabase. The Street Category is now listed as PAVED.

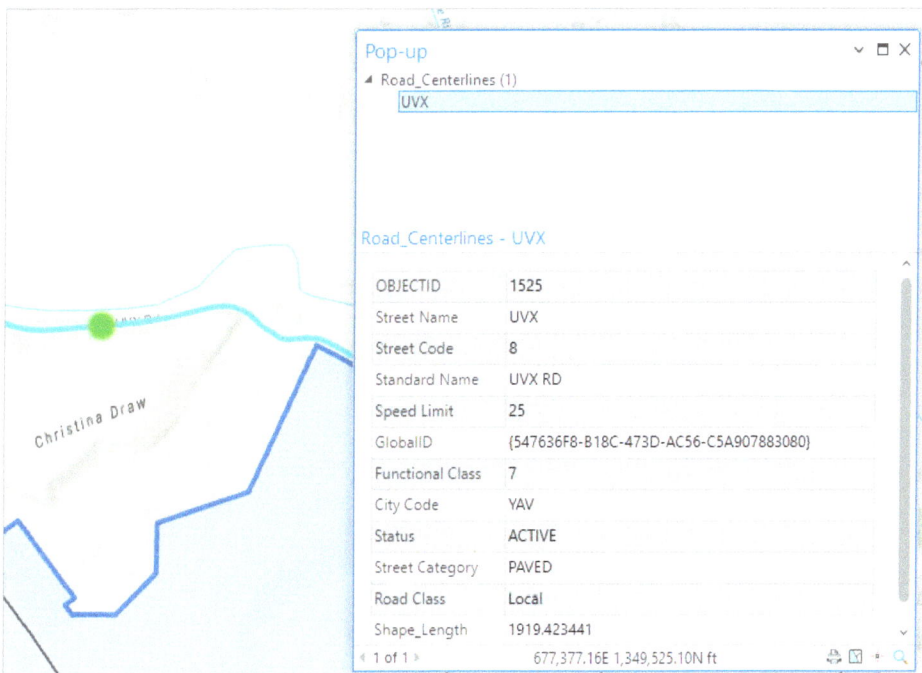

This exemplifies the power of ArcGIS Enterprise, where all components are interconnected, ensuring data consistency and real-time updates across the entire system.

Leave the organization

The editing and sync process is performed multiple times until all the road segments near the school locations have been paved. When the project is complete, the collaboration between these two organizations ends. In this final part of the tutorial, you will perform the steps needed for the guest to leave the collaboration and for the host to delete the collaboration.

1. Return to the ArcGIS Enterprise browser tab and navigate to **Collaborations**.

2. Find the **Roadwork project** collaboration. Click the **Action** icon and click **Leave Collaboration**. Confirm by clicking **Leave Collaboration**.

   The guest organization has successfully left the collaboration. There are no listed active collaborations.

Collaboration Name	▲ Participation	Creation Date	Action
	No collaborations found.		

3. Navigate to the ArcGIS Online browser tab and go to **Collaborations**.

4. Find the **Roadwork project** collaboration. Click the **Action** icon and click **Delete Collaboration**. Confirm by clicking **Delete Collaboration**.

   The collaboration was deleted successfully.

   On the guest side, both the ArcGIS Online group and the **Medio Public Assets** web feature layer are not deleted after ending the collaboration. However, since the project has ended and the organization no longer needs the data, it will delete the group and the data.

5. On the navigation bar, click the **Groups** tab.

6. Open **Medio County Group** and go to the **Settings** tab.

7. At the top of the page, disable the delete protection.

8. Then, click **Delete group**.

   The group is successfully deleted. However, the **Medio Public Assets** web layer remains.

9. Go to the **Contents** tab and check the box for the **Medio Public Assets** web layer.

10. Click the **Delete** option.

   With these steps, you have successfully deleted the data from the **Roadwork project**.

Summary

In this chapter, you successfully collaborated with a local road company to begin work on unpaved roads, using the dashboard created in the previous tutorial. By establishing a distributed collaboration between your ArcGIS Enterprise organization and Medio Contractors LLC's ArcGIS Online account, you enabled seamless data sharing and editing. You took on the roles of both the host and the guest organization to fully understand the process of sharing data through a distributed collaboration. You also confirmed firsthand that synchronized edits are reflected throughout the entire ArcGIS ecosystem, including dashboards, web maps, web feature layers, and the source enterprise geodatabase.

In this section of the book, you improved your department's data accessibility by publishing a web feature layer to your organization ArcGIS Enterprise portal. You then tested the web-editing experience by making attribute and geometry changes using Map Viewer. You were then tasked to assess the road conditions around the school locations for which you performed a spatial analysis. Toward the end, you presented your analysis to the entire department. To do that, you conveyed the results of the spatial analysis in a more interactive and easy-to-read format for the public using ArcGIS Dashboards. Finally, you collaborated with a local firm to act on the roadwork. To share data between an ArcGIS Enterprise and an ArcGIS Online organization, you created a distributed collaboration. In essence, through these five tutorials in part 3, you created a comprehensive authoritative workflow by using the power of ArcGIS Enterprise.

15

Maintaining and troubleshooting in ArcGIS Enterprise

THE FINAL PART OF THIS BOOK FOCUSES ON HOW TO MAINTAIN and troubleshoot in ArcGIS Enterprise. These last five chapters will address topics related to performance best practices, troubleshooting basics, patching ArcGIS Enterprise, creating and using backups, and how to upgrade ArcGIS Enterprise. These chapters tie together the subject matter of this book by demonstrating how to keep your ArcGIS Enterprise organization stable and performing well, while at the same time providing you with the tools to investigate, resolve, and recover from different issues. By the end of part 4, you will understand the basics of a sound backup strategy and how to maintain your organization over time.

CHAPTER 16

Performance best practices

Objectives

- Optimize layers before publication.
- Optimize service instances.
- Monitor service performance.

Introduction

Performance and scalability is an important pillar of a well-architected ArcGIS Enterprise deployment. In this chapter, you'll learn how to ensure that users of ArcGIS Enterprise have a good experience accessing the resources they need. Because users will be accessing those resources over the network, they may experience slower response times than when working with local data. For that reason, speed optimizations are an important consideration, and you'll learn how to tune services to make them as fast as possible.

But speed isn't the only consideration. GIS resources that users access through ArcGIS Enterprise must meet their needs. To use any of the optimization strategies in this chapter, you must have a strong understanding of what those needs are. You will need to talk to users of the software and find out what capabilities they require and not rely on assumptions about what will work for them.

Optimize layers before publication

Making sure your services work as well as possible starts before you even create the service, by configuring the layer before publication. You can implement several strategies to improve the performance of services.

Generalize geometry

For vector data, every vertex of every visible feature gets drawn. More complex geometries with more vertices take longer to render. For many use cases, the data may contain far more vertices than required. You can simplify the geometry, reduce the number of vertices, and substantially speed up rendering time.

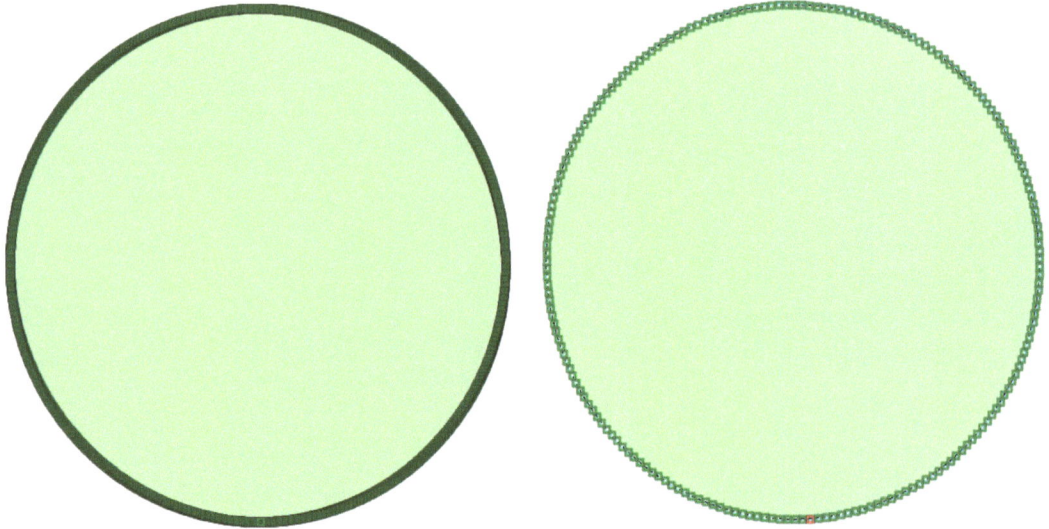

Figure 16.1. The feature on the left has 2,027 vertices. The feature on the right has 203 vertices. At most scales of visualization and analysis, the additional vertices on the left won't make any meaningful difference in usability but will be substantially slower to process than the feature on the right.

Simplify cartography

Complex symbology with multilayered symbols, color gradients, and textures creates additional computational complexity when rendered for display. Complex rules for label placement can likewise make it more difficult to render the data in a map.

Sometimes, those complexities are required to enable the work users must do with the service. But because sophisticated cartographic representations slow a service, you should weigh their importance before publication. You might also consider creating multiple layers with different representations to meet some users' needs for complex cartography while meeting other users' needs for faster rendering.

Use visibility range

When the map view is zoomed out, rendering everything is often a bad idea for two reasons: It makes the map hard to read, and it makes the service slow to respond. Both problems are related to having to render a large absolute number of features and labels. One way to fix both problems is to vary the features that are displayed by the scale of the current map view. Only the most important features are rendered when zoomed out, but as you zoom in, you can display more features.

When you use scale visibility, any given map view draws a more limited number of features. There is sufficient space to render each feature, which makes it easier to read the map. And with fewer features being drawn, they display faster in the map.

Figure 16.2. With scale visibility of features, the smaller scale map of Sierra Leone on the left shows only national boundaries, major highways, and capital cities, whereas the larger scale map on the right shows more detailed information. Courtesy of Esri and USGS.

16

Filter data

Some datasets have more features than are needed for their purposes. This might be due to including features outside the area of interest or because some features don't have the relevant attributes. When these features are included in the published service, the service is slower to render and harder to use than if the irrelevant features were excluded. For that reason, you may want to filter out those features before publication.

A dataset may also include many more attribute fields than necessary. This tends to have less effect on drawing performance than including additional features, but it has a strong impact on usability. Users do not want to scroll past 100 fields to find the attribute they are looking for. By limiting the number of fields, you can make the service much easier to use while simultaneously reducing the amount of data that has to be transmitted.

Choosing the right type of service to publish

Chapters 9 and 10 covered the options available for publishing different types of services. That choice affects how well services perform, so you want to make the right decision to ensure users of each published service have an optimal experience.

Hosted versus nonhosted services

Hosted services are optimized to reduce the amount of memory used for each service. In situations in which you need to publish many services, hosted services are a good option to prevent overtaxing the memory of the machine where ArcGIS Server is running.

The optimization of hosted services, however, means that they are not as user-configurable as nonhosted services. Hosted services also have the limitation of being able to read data from only specially configured locations. Nonhosted services provide greater flexibility and control over exactly how the service is configured. If you require a service to read data directly from a user-managed data store, you must use a nonhosted service.

Hosted and nonhosted services are also distinguished by the underlying technology that enables the service to respond to requests. On each machine where ArcGIS Server is running, a single process responds to all requests for all hosted services. Nonhosted services, on the other hand, are enabled by multiple processes. The processes that respond to requests on nonhosted services are called service instances but are also referred to as ArcSOC. In a functioning site, you can expect to see a number of ArcSOC.exe instances running to power these services.

Because the underlying technologies of hosted and nonhosted services are different, you cannot easily transition a service from one type to the other. That's important because some use cases require a particular type of service. Versioned editing workflows, for example, require a nonhosted service. Be aware of how a service will be used when you publish it, so that you can make the right choice between a hosted and nonhosted service.

The different technologies of hosted versus nonhosted services also affect the appropriate strategy for scaling services if system resource constraints are causing them to perform suboptimally. Because a single process per machine responds to all requests on all hosted services, the appropriate scaling strategy for hosted services is to add more system resources to the ArcGIS Server site. You can do that by either adding additional system resources to the machine where ArcGIS Server is running or adding more machines to the ArcGIS Server site. For nonhosted services, however, you can increase the number of service instances to provide more resources to that service without increasing the amount of resources for the entire ArcGIS Server site.

Map service versus feature service

The main difference between a map service and a feature service is the application that renders geospatial data when a client makes a request to ArcGIS Server. In a map service, the data is rendered by ArcGIS Server, and the client that made the request receives an image of the data. In a feature service, ArcGIS Server responds with the geometry of the features, which is rendered by the client that made the request.

Whether data is rendered server-side or client-side has several important implications for performance:

- Only feature services enable editing, because map services do not provide the vertex geometry to the client.
- Feature service responses are typically larger in size than map service responses, increasing the network bandwidth required.
- Map services can support the same advanced cartography as ArcGIS Pro, regardless of the client. Feature services need to have less sophisticated symbology that can be rendered by any client.
- Feature services can include only vector data, whereas map services can include raster and vector data.
- Feature services place more computational demand on the client application, whereas map services place more computational demand on ArcGIS Server.
- Because of the different advantages and limitations of each type of service, it can be desirable to make data available as both a map service and a feature service. For example, you can make the feature service available to users who must be able to edit the data. Simultaneously, you can make the map service available to users of low-powered devices to reduce the client-side processing required.
- The process for creating a map and feature service for the same data depends on how the service is published. For nonhosted services that reference user-managed data, you will always get a map service. You can turn on feature access to create a nonhosted feature service that references the same data. If you have a hosted feature service, you can use ArcGIS Pro to publish a hosted map service that references the same data.

16

Data versus tiles

Whether they are hosted or nonhosted, both map and feature services are data services. The service maintains a connection to the underlying dataset. That connection enables users to run queries against the service to return only a subset of data. Alternatively, geospatial data might be made available as a tile layer. A tile layer is not directly connected with the underlying data and responds to requests with precached tiles.

A tile layer has the advantage of responding more quickly to requests than a data service. Lacking a direct connection to the underlying data, however, means that you cannot query data from a tile layer, and it does not reflect edits made to the data since the last time the cached tiles were created. In situations where data is updated infrequently and you do not need to query that data, a tile layer is better than a map or feature service. Tile layers are an especially good choice for basemap layers.

When creating a tile layer, there are two options: raster tile and vector tile. You can use a raster tile for any kind of data, but you can use a vector tile only if all the underlying data is in vector format. Vector tile layers are generally better because they are faster, responsive to the screen size on the client device, and can be restyled. Use a raster tile layer only when you cannot use a vector tile layer.

Configure services

Although you cannot easily change the fundamental nature of service after publication (such as by converting it from hosted to nonhosted), you can make many other changes to the service configuration that can improve performance.

Service capabilities

Because a service defines what users are allowed to do with the underlying data, the capabilities enabled in the service have a substantial effect on how usable the service is. For example, a service might have editing capabilities turned off. That would be appropriate for a service that should be read-only, but it's a problem if users need to make changes to the data through the service.

For hosted services, you can configure the editing and export capabilities. Nonhosted services have the potential to configure additional capabilities, such as

- network analysis,
- version management,
- linear referencing,
- parcel fabric,
- utility network,
- additional data formats (WCS, WFS, WMS, KML, OGC features).

Fictional user story

Medio County wants to make its parcel data more accessible to the public and has decided to publish the data from its enterprise geodatabase as web layers to ArcGIS Enterprise. Elise Medina, the county GIS manager, has been meeting with both internal stakeholders and members of the community to find out what they need from these services. Based on those meetings, she has collected this list of requirements:

- Property owners need to be able to get information about their own property, such as the legal description, property ID, and tax owed. Many property owners live in parts of the county with poor internet service or access the internet primarily on mobile devices.
- Real estate firms need to get information about any property in the county so they can do market analysis.
- Clerks from the county assessor's office need versioned editing to enable multiple users to edit parcels simultaneously.
- Private developers need existing parcels to use as a basemap for overlaying new development plans for comparison. The data reflected on the basemap needs to be no more than six months old to fit the developers' needs.

From this list of requirements, Elise decides to create three services from the parcel data:

- A publicly shared nonhosted map service referencing the data in the registered user-managed enterprise geodatabase. This will enable property owners and real estate firms to query the parcel data, while keeping the rendering server-side to assist with poor internet service or low-powered devices.
- A nonhosted feature service created by enabling feature access on the map service. This service is shared only within the Medio County ArcGIS Enterprise organization. That enables versioned editing for the assessor's office, without exposing the service publicly.
- A publicly shared vector tile layer, which will allow developers to quickly visualize the parcels as a basemap layer. Elise creates a workflow to update the vector tile layer every quarter to meet the developers' needs for relatively recent data.

When you configure services, ensure that you enable all the capabilities that users will need. For performance and security reasons, it's also a good idea to disable capabilities that users won't need.

16

Shared instances versus dedicated instances

For nonhosted services, you can choose between shared and dedicated service instances. Dedicated instances are ArcSOC processes that respond to requests for a single service. Shared instances are a pool of ArcSOC processes in which any ArcSOC in the pool can respond to requests for any service configured to use shared instances.

Because a service instance consumes memory even when it is not actively handling requests, dedicated instances are usually less memory efficient than shared instances. This is especially true for services that don't receive many requests so dedicated instances will sit doing nothing most of the time. If that service uses the shared instance pool instead, the ArcSOCs in the pool will respond to other services instead of sitting idle.

The memory efficiency of shared instances means you should use them when you can. But not every type of service can be configured to use the shared instance pool. Map and Image Services (with or without feature access enabled) that were published from ArcGIS Pro can use shared instances, but other types of services must use dedicated instances. Shared instances don't support every capability of a map service either. Notably, using versioned editing capabilities requires dedicated instances.

Because choosing between shared and dedicated instances can be configured after publication, you can easily move a service from one type to the other. That's useful in cases in which the original publisher chose the incorrect type. It's also helpful in cases in which the usage of a service changes over time. For example, if a service level agreement (SLA) that required dedicated instances is no longer in effect, you can migrate that service to use shared instances without affecting the service's capabilities.

Configure the number of instances

Each ArcSOC process can handle a single request at a time. Generally, an ArcSOC will respond to a single request in less than one second, but the responses aren't instant. If a service receives more requests per second than can be handled by a single ArcSOC, the response time will become increasingly slower because requests need to wait their turn.

For nonhosted services that use dedicated instances, you can configure the minimum and maximum number of service instances. If the current number of ArcSOCs cannot keep up with the requests, ArcGIS Server will create additional ArcSOC processes until it reaches the maximum allowed. ArcGIS Server will terminate ArcSOCs above the minimum if they sit idle for too long.

Creating a new ArcSOC typically takes less than one second, but it isn't instant either. Consequently, you want to set your minimum number of instances high enough so that users don't commonly experience a "cold start," in which the system needs to create new ArcSOCs to keep up. The maximum number of instances should be set high enough to accommodate peak demand on the service. The number of running instances also needs to be balanced against available machine resources, since every ArcSOC uses memory, even when it isn't actively responding to requests.

There isn't a scientific way to predict exactly how many instances you will need for a given service. And that need may change over time anyway. For that reason, it's not important to get the number of instances exactly right when you first create a service. Instead, monitor the service responses and adjust accordingly.

Caching map services

To speed up responses on map services, you can choose to create a cache for that service. Instead of spending time redrawing the response for every request, the responses can be predrawn. Cached map services are therefore much faster to respond to requests than dynamic map services.

Cached map services are similar to tile layers in that they respond with predrawn images. Like tile layers, cached map services won't reflect any edits made to the data since the cache was created, so they work best for data that is infrequently updated. But cached map services are different from tile layers in that they maintain a connection to the underlying data. You can still submit a query on a cached map service and get attributes for the features in the map.

Caching hosted feature layers

When ArcGIS Enterprise is configured with an object store, you can cache responses to requests on hosted feature layers. This type of caching is different from creating a cached map service because the responses aren't pregenerated. Instead, the first time a user submits a request on the layer, ArcGIS Server generates the response as normal but the response is also saved in the object store. The next time a user makes that same request, they get the cached response instead of having ArcGIS Server regenerate the same response again.

Configuring response caching for hosted feature layers is a good idea if you have users that are making many of the same requests. Response caching also works best with data that doesn't change frequently, because edits will cause the cache to be regenerated, which reduces the benefit of caching responses.

16

Fictional user story

Jim Yazzie, the GIS expert at the Becken Pond Conservation Society, first set up ArcGIS Enterprise back in 2018, when 10.6 was the current version. He's kept it updated since then, but all BPCS nonhosted services are configured for dedicated instances, which was the only option at 10.6 and hasn't been changed since.

The system resources of the machine where ArcGIS Server is installed are currently fully used. An important project can't move forward without publishing additional data as a new nonhosted service, but there aren't enough system resources to do so. Linda Jackson, the BPCS executive director, wants Jim to present some options for getting the data published.

Jim sees three possibilities:

- Increase the system resources available to the ArcGIS Server site. This is Jim's preferred option because he would like to add another machine to the Server site to enable greater resilience in addition to greater service capacity. This option has the downside of additional expenses.
- Move the less commonly used services to use zero minimum instances. This option has the disadvantage of making services slower to respond because of the need to create a new ArcSOC to handle even a single request.
- Move eligible services to use the shared instance pool.

Linda vetoes the first option for budgetary reasons. Jim decides that a combination of the other two options can free up enough system resources. He turns on shared instances for every service that can use it. For a handful of infrequently accessed geoprocessing services that can't use shared instances, he sets the minimum instances to zero.

Monitor service performance

Another key pillar of a well-architected system is observability. You need to be able to gather information about the current state of your system to verify that the system is in a heathy state and to enable rapid response to any issues that occur.

Server statistics

ArcGIS Server comes with the ability to configure reports on the services published to a Server site.

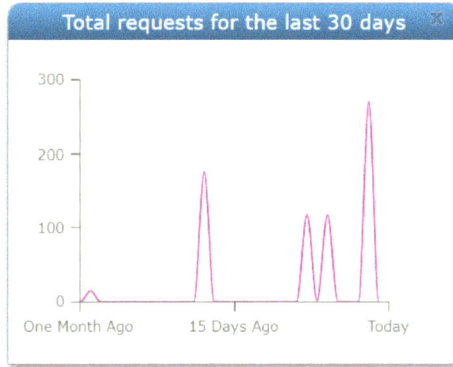

Figure 16.3. Example of a server statistics report.

Each report is displayed in a dashboard view of charts, and the data can be exported for further analysis. You can configure a report for a single service or an aggregation of services. Server statistics reports show metrics over a configured time period and can provide the following information:

- Total number of requests made to measure the load on the service
- Average response time to measure the typical length of time a client had to wait for a response
- Maximum response time to measure the worst-case length of time a client had to wait for a response
- Timeouts, to measure how many requests did not receive a response due to timeouts
- Service maximum running instances to measure how many ArcSOCs were active and available to respond to requests

Server statistics are useful for responding to trends over time. For example, if the maximum response time is trending upward because of increasing demand on a service, you can increase the number of instances for that service before users notice any performance degradation.

16

Logging for performance

In addition to statistics, ArcGIS Server also captures logs of service events. If you have a system to capture and report on events in the ArcGIS Server logs, you can use those logs proactively to address a potential issue before it becomes a problem. More commonly, however, logs are used reactively for troubleshooting a problem that has already occurred.

There are four main steps when using ArcGIS Server logs for troubleshooting performance issues:

1. **Increase the logging level.** By default, ArcGIS Server logs only error and warning events. Those events may not provide the level of detail about the problem you need to solve. Increasing the logging level won't retroactively make old events appear in the logs, but it will start logging events at greater detail going forward.
2. **Replicate the problematic behavior.** Replication confirms that the problem exists and ensures the logs capture events associated with the issue.
3. **Review the logs for information that can explain the issue.** For example, if the issue is related to slow responses on a service, you might look for those log events that took an inordinate amount of time to complete. In addition to looking up the event descriptions, you can also look up the log codes to get more information about an event. This is the most challenging and time-consuming step, and you may find it helpful to use specific log analysis tools, such as the System Log Parser created by Esri's Professional Services division.
4. **Set the logging level back to its original configuration after the issue is addressed.** Highly detailed logs are useful for finding the source of problems, but recording all those events requires system resources and can quickly use up substantial amounts of disk space. For that reason, you don't want to capture detailed logs all the time.

ArcGIS Monitor

One disadvantage of ArcGIS Server logs and statistics is that they capture information related to only a single ArcGIS Server site. Gathering information across multiple server sites, or information related to other components of the ArcGIS system (such as the network or databases), requires collating information across multiple sources.

ArcGIS Monitor is one way to integrate information from a variety of sources to better analyze your ArcGIS Enterprise system as a whole.

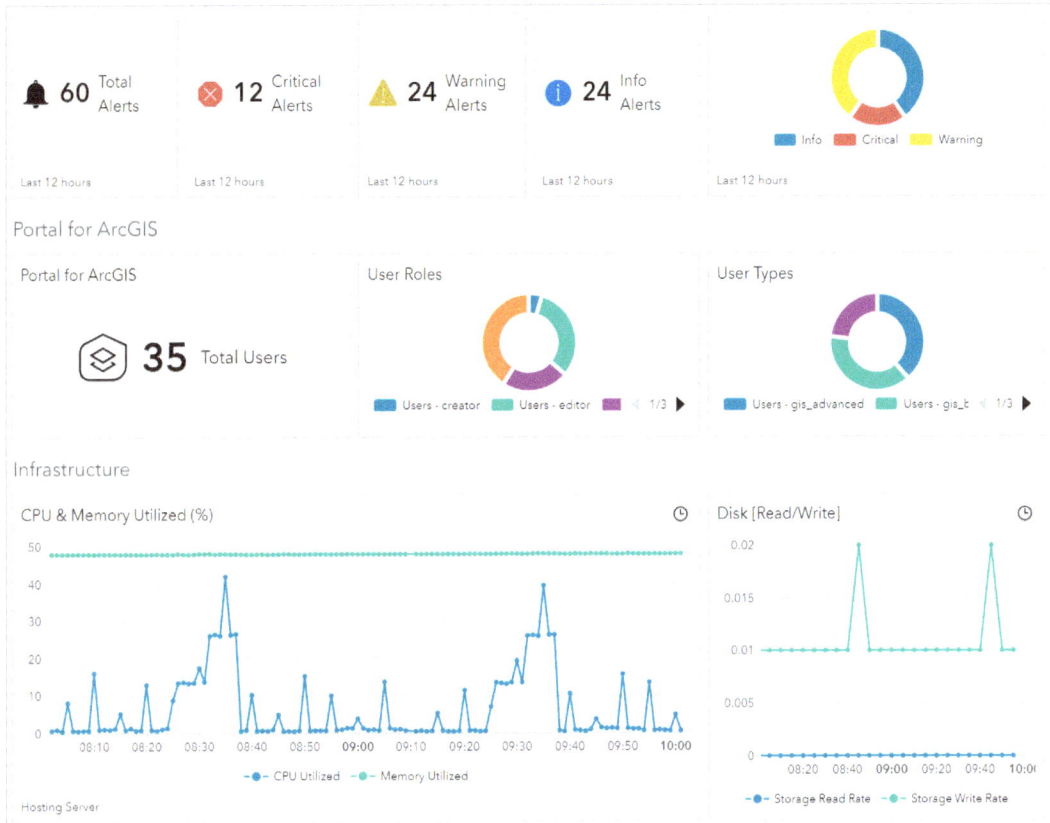

Figure 16.4. ArcGIS Monitor brings information from many components into a single application for alerting and analysis.

You can use Monitor to configure alerts to let you know when part of the system has exceeded the threshold for what you consider healthy behavior. Alerts are helpful for making you aware of issues right away, without having to wait for reports from users that there is a problem. For example, a good rule of thumb is to not consistently use more than 80% of the machine's memory, so you might configure an alert when memory goes above 80%. You can investigate whether the cause of the increase was some transient problem or if it is indicative of a larger issue that you need to address.

Monitor's integration of information can also help you when working reactively in response to a reported issue. For example, you may get a report from users about a geoprocessing service that is slow to respond. In this case, perhaps the problem is caused by insufficient CPU on the ArcGIS Server machine, but users would have no way of knowing that. You could use either Monitor or the built-in server statistics to confirm the slower response time, but that wouldn't help you understand why it is slower. Monitor could give you additional information about the corresponding increase in CPU usage that is happening at the same time as the slow service responses. Without that integration of data sources, it would be more challenging to discover the root cause of the problem.

Fictional user story

MegaBiz has a large and complex deployment of ArcGIS Enterprise. It has eight ArcGIS Server sites, and most of those sites have multiple machines joined to them. They also have an available ArcGIS Enterprise, as well as multiple database and web app server machines. In total, they have 36 machines involved in their deployment.

Recently, the machine where MegaBiz has its ArcGIS-managed raster data store experienced an intermittent failure of one of its solid state drives (SSDs). This problem originally manifested as users who were unable to run raster analysis jobs using MegaBiz's ArcGIS Image Server site dedicated to raster analytics, which is installed on a different machine from the raster data store.

Viktor Dudnyk, the GIS systems administrator responsible for analytics services, first checked the server statistics and couldn't see anything unusual in terms of timeouts, high number of service requests, or slow response times. He could replicate the issue, but he couldn't gather enough information from the logs to understand why the analysis request was failing.

In the meantime, users also began reporting problems with viewing hosted image services published to the Image Server site dedicated to image hosting. This is a separate site from the raster analytics site, but both use the same raster data store, which was the underlying cause of both problems. The issue reports for this problem get routed to Yvonne Akindele, the GIS systems administrator responsible for read-only hosted services. Just like Viktor, she has trouble identifying the source of the problem from the server logs and statistics for the image-hosting ArcGIS Server site.

Eventually, Viktor and Yvonne realize that their problems have the same cause. They fix the issue by replacing the failing SSD, but not before some costly downtime.

In the aftermath, Harjeet Singh, the MegaBiz director of data, wants to know what could be done to minimize downtime in the future. Viktor points out that the complexity of the system and fragmentation of relevant logs and responsibilities make it challenging to find the source of problems. Yvonne suggests that it would be appropriate to install Monitor to put all the information from across the system into a single application. She notes that if it had been installed, they could have configured alerts to notify them of the failing SSD before users even noticed, averting the problem in the first place.

Tutorial 16: Investigate system performance

In this tutorial, you'll first examine a machine where ArcGIS Server is running to ensure that it has sufficient resources for your services. Then you'll use server statistics to investigate the performance of services over time. Lastly, you'll examine the configuration of services to identify opportunities for optimization.

Examine machine resource usage

1. Sign in to a machine where ArcGIS Server is running.

2. Using a system management tool that reports on resource usage (such as Resource Monitor on Windows or htop on Linux), answer the following questions:
   - How much memory (RAM) does the machine have?
   - What percentage of memory is currently being used?
   - How many CPU cores does the machine have?
   - What percentage of available CPU is currently being used?
   - How much disk space does the machine have?
   - What percentage of disk space is currently being used?

3. Using a system management tool that reports on individual running processes (such as Task Manager on Windows or htop on Linux), answer the following questions:
   - How many ArcSOC processes are currently running?
   - Generally, how much memory is each ArcSOC process consuming?
   - Given the available memory on the machine and the amount of memory generally consumed by a single ArcSOC process, estimate how many additional ArcSOCs you could create before you start going over 80% memory usage.
   - Does this machine have sufficient resources for your current and planned services?

Examine server statistics

1. In a browser, open the Server Manager web app for an ArcGIS Server site federated with ArcGIS Enterprise that you have administrator privileges for.

2. In the upper right, click the **Logs** tab.

3. Click the **Statistics** tab and then click **New Report**.

4. On the left, check the box next to one of your services to report statistics for that service only.

5. For **Show**, select **Maximum Response Time**.

6. For **Age**, select **Last 30 days**. Then answer the following questions:
   - Have the response times been acceptable?
   - Can you identify any patterns, such as cycles or trends, in response times?
   - Is there an upward trend in response times that if left unaddressed might become unacceptable in the future?

Examine hosted and nonhosted services

1. In the Server Manager, on the top, click the **Services** tab.

2. Review the list of services published in the **Site (root)** folder, which is the default view.

3. On the left, if you have additional directories beyond those labeled **System** and **Utilities**, review the services in those directories.

4. If there is a directory labeled **Hosted,** click it. If not, you have no hosted services published to this server site and can skip to step 6.

5. Review the list of hosted services.

6. Answer the following questions about the hosted and nonhosted services published to this ArcGIS Server site:
   - Are there any nonhosted services that could be optimized by republishing them as hosted services (for example, an infrequently used service that does not need to reference user-managed data)?
   - Are there any hosted services that could be optimized by republishing them as non-hosted services (for example, a service that should have the version management capability)?

Examine nonhosted service instance configuration

1. In the **System** or **Utilities** folder, click the name of one of the nonhosted map services.

2. On the left, click the **Capabilities** tab and answer the following questions:
   - Which capabilities are configured for this service?
   - Are all the required capabilities enabled?
   - Are unneeded capabilities disabled?

3. Click the **Pooling** tab and answer the following questions:
   - Could this service be configured to use shared instances (if so, will there be an option to toggle between shared and dedicated instances)?
   - Does this service currently use shared instances?
   - If the service uses dedicated instances, what are the minimum and maximum number of instances configured for the service?
   - Considering how the service is used, are the instance type and number configured correctly?

4. Click the **Caching** tab and answer the following questions:
   - Is this map service drawn dynamically from the data or using tiles from a cache?
   - Is the caching option set appropriately for this service, considering how it is used?

Take the next step
Change the configuration of services that could be further optimized.

Summary
In this chapter, you learned how to optimize your services and investigated your ArcGIS Enterprise deployment for opportunities to ensure services are configured to provide users with the best possible experience.

16

CHAPTER 17

Troubleshooting in ArcGIS Enterprise

Objectives

- Create an action plan to efficiently react to an unexpected issue.
- Learn the logging capabilities of ArcGIS Enterprise.
- Investigate behaviors through web tools.

Introduction

ArcGIS Enterprise can create many data connections to support mobile workers, offline GIS analysis, and map creation. Because of the level of complexity of these workflows, users may sometimes run into unexpected behaviors. Other factors, such as network changes, improper workflow design and application, or defective behavior, may cause users to experience unexpected issues in their day-to-day use of ArcGIS Enterprise.

This chapter will serve as a primer on how to gather information about an issue, understand the scope of the problem, and communicate the issue to key stakeholders. After establishing a base of understanding, we will also cover what types of triaging and troubleshooting methods are available to you as an administrator. Although this chapter will not cover every troubleshooting scenario, we will cover these basics:

- Establishing system history at the time of the error
- Gathering and reviewing logs from relevant locations
- Initial triaging of the error and checks on basic system functions

This chapter will also discuss how to work with Esri Technical Support, highlighting how to collaborate with support analysts and specialists and provide them with data that is relevant to your issue.

The chapter will conclude with an examination of tooling that's been created to quickly gain key insights into your configuration of ArcGIS Enterprise and an exploration of alternative logging locations through the ArcGIS Enterprise portal administrative end point—and the ArcGIS Server administrative end point.

What to do when you encounter a technical issue

Throughout this book, we've introduced ArcGIS Enterprise as a system of federated software components to achieve a specific goal. In this system, many users find they can apply more than one solution to their use case. As solutions become complex or changes require iterations, you may find yourself running into different issues or workflow limitations. Other issues may be caused by external factors or defects within the software. Although the origins of these issues are diverse, an administrator can take similar steps to get more information about those issues.

How an ArcGIS Enterprise administrator deals with the unexpected and plans for the unknown makes the difference in the overall recovery time of the deployment. The most important thing to remember is not to panic or make major changes to the environment that are not completely understood. For example, in a scenario in which a high availability (HA) deployment is not properly promoting to a standby portal, you may be tempted to apply a backup to the failed machine and spin it back up. However, if the failure in the primary/passive switch is not understood, there is a chance that the deployment may be caught in a "split brain" state, with each Portal for ArcGIS instance believing that it's the primary instance. This would lead to a more complicated solution that may add time to the overall recovery.

So, what should you do if you encounter an issue? Although each issue is different, let's talk about what steps could be taken for an effective initial response.

Observing the issue

It may seem obvious, but the first step in solving the problem is understanding what the problem is. Getting on a call with the user and observing the problem is crucial here. For example, if a user is trying to edit a feature in ArcGIS Field Maps, get the basic information on what piece of content or service the user is trying to work with, along with what action is causing the problem. When you have the basic details, you can try to reproduce the problem on your end to make the investigation easier. We will dive deeper into how to investigate an issue later in this chapter.

Isolating the issue

In Esri Technical Support, analysts are trained on taking a problem as it is presented and paring it down to its most basic component. If a user is unable to edit a feature layer in Field Maps, you may ask if editing is possible in Map Viewer. If you are unable to edit the service in Map Viewer, you have removed Field Map's relevancy from the equation, making troubleshooting slightly easier. Through a series of questions and logical answers, the exact location of the failure can be identified.

Categorizing severity

Through your investigation and isolation efforts, you should begin forming an idea of how disruptive the issue may be. Continuing with our example, let's begin by investigating the Field Maps issue as it presents, meaning it directly affects one user. However, as you continue to investigate, you learn that multiple services from the same data store are also affected, expanding the pool of affected users to a working group. Your perspective on the severity of the issue should be reported because it is a valuable nuance to those who may help you resolve the issue.

Communicating the issue

Combining your understanding of the issue, isolating it to a particular workflow or group of users, and categorizing its severity, an ArcGIS Enterprise administrator should begin to communicate the problem to their key stakeholders. All affected users, other ArcGIS Enterprise administrators, and IT team members need to be briefed on the details of the issue. Setting an informational banner and updating it with your progress is a great idea to keep your user base informed of what is happening. If the issue is severe enough, other measures may be taken to limit changes happening to the environment as troubleshooting takes place, such as setting ArcGIS Enterprise to read-only mode.

> Tip: Esri is committed to ensuring that ArcGIS Enterprise users are supported in whatever system configuration is chosen. Esri Technical Support is composed of industry experts on the ArcGIS platform. If you feel as if you need support during any of these steps, and you have maintenance coverage on ArcGIS Enterprise, you may call our staff to get more help. We will cover more on what information is helpful to send to technical support later in the chapter.

17

These steps serve as a basic example of how you may start to respond to an issue. As you gain more experience, you may find yourself amending these steps to match your response style. As you go through more opportunities to troubleshoot issues, you will learn more about how to effectively troubleshoot and where to look for key information.

Investigating an issue in ArcGIS Enterprise through logs

We've talked about establishing an effective first response against a possible issue. Although observing and isolating the issue down to its core will reveal a lot of details about the problem, there are additional ways to gather more actionable information about the observed behavior. Because errors can occur at various levels in ArcGIS Enterprise and disrupt different functions and features throughout, it is best to remain open to new troubleshooting paths and ideas. The best place to start is to make sure you can reproduce the error on demand.

Reproducing the issue, or triggering the issue on demand, is a dependable way to generate different pieces of information as you begin to troubleshoot. By having a process to make the problem happen repeatedly, you can generate logs, network traffic, and other pieces of information that may be time sensitive. If you decide to work on the issue with technical support, reproducing the issue on demand makes the troubleshooting process more efficient, since time will be saved identifying a trigger. Once we've established a surefire way of causing the issue to recur, we can move on to gathering information and isolating the issue.

Before considering making any changes to your ArcGIS Enterprise deployment, you need to understand what is happening in your environment before, during, and after the issue occurs. ArcGIS Enterprise contains built-in logging capabilities at each of its major components: Portal for ArcGIS, ArcGIS Data Store, and ArcGIS Server. Logging in the ArcGIS Web Adaptor is primarily handled through the web server it's deployed on (such as Microsoft's IIS or the Java platform for Linux). Logging in these three components may look different, depending on how they are accessed, but the following features are consistent:

- **Logging level**: The level of details that logs deliver is controllable through the Log Level Selector. In order of importance these are **Severe**, **Warning**, **Info**, **Fine**, and **Debug**. Outside of actively troubleshooting an issue, users are advised to set logs to **Warning** or **Severe**; leaving the levels any finer may result in negative system performance.
- **Filtering**: Logs can be filtered by different objects within each component. For example, ArcGIS Enterprise portal logs can be filtered by different users, whereas ArcGIS Server can be filtered by the service in question.
- **Storge and retention**: Administrators can set the location where the raw log files are stored as well as the retention policy for these files. This aids in preserving system resources.

As logging levels become finer and finer, the amount of information that is captured in different parts of ArcGIS Enterprise increases rapidly. Although using a finer log level is necessary during active troubleshooting, leaving logging at a finer level for a long period of time will lead to more space being used up to store these log entries and impact performance within ArcGIS Enterprise. The next graphic demonstrates how many logs are captured during a failed publishing job in the same time period.

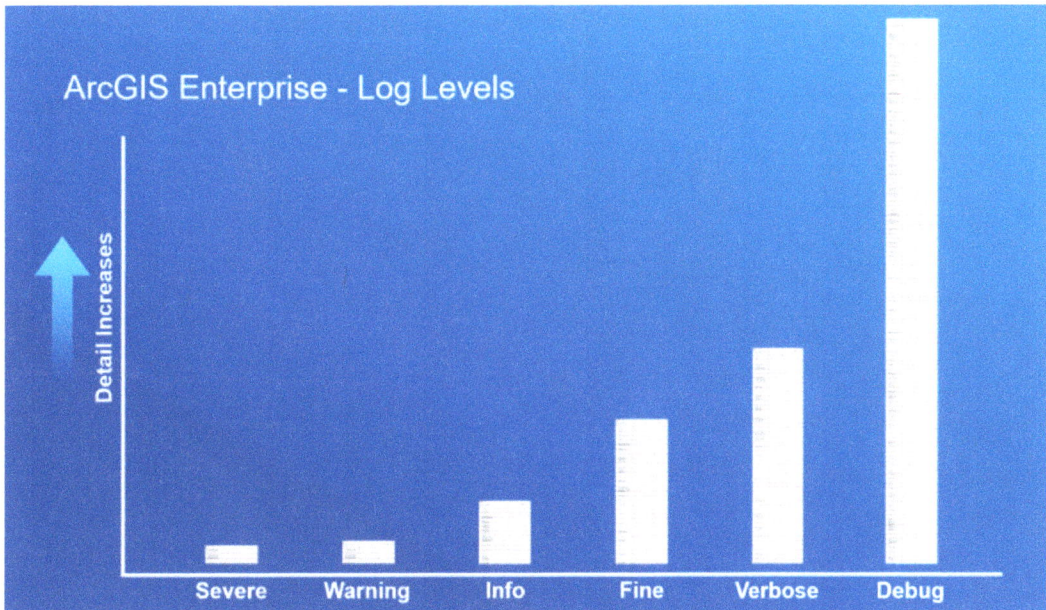

Figure 17.1. After a failed publishing task, the logging level at which you set your ArcGIS Enterprise deployment determines how detailed the portal log will be.

Now that we've discussed some commonalities of logs, let's review where logs can be accessed and in what context they should be used.

17

ArcGIS Enterprise portal logs

Log Settings

The portal is currently logging events at the **WARNING** level.

To change the level of detail at which portal logs events, click Settings.

Query Log Messages

Query log messages by selecting from the following options, and clicking Query.

Log Level:	WARNING ⌄
Source:	All ⌄
Start Time:	[_____]* End Time: [_____]*
	Example: 2015-01-31T15:00:00 Example: 2014-12-31T09:00:00
Log Codes:	[_____]*
	Example: 200000-201999, 207026
Users:	[_____]*
	Example: admin, portaladmin
RequestIDs:	[_____]*
	Example: 7b85af92-84eb-400d-b28d-ef9e704358ad
Message Count:	1000 *
Federated Servers:	None ⌄
	*Optional

Format: HTML ⌄

Query

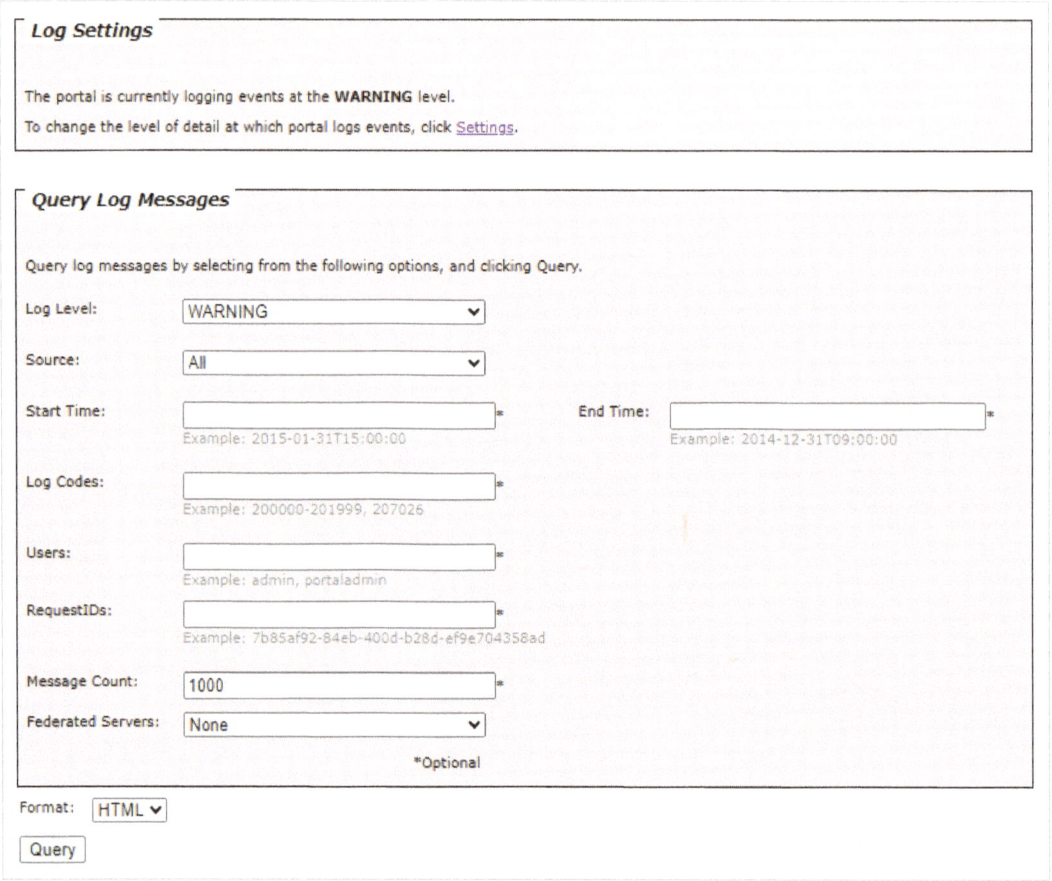

Figure 17.2. You can adjust the level of details for your portal logs and query previous log messages.

The ArcGIS Enterprise portal logs can be accessed by administrators in the portal adminis-trative end point. General portal logs are accessed by selecting Logs at this end point, which display settings and filters of logs available for analysis.

These logs are best used when troubleshooting organization-wide issues. If you are trying to troubleshoot users being able to sign in, certain built-in applications not working cor-rectly, or issues connecting to hosted feature layers, these logs are a good place to start.

> *Note: If troubleshooting issues with specific ArcGIS Enterprise features, such as distributed collabo-rations, logging for these features is located within the portal administrative end point as well.*

ArcGIS Server

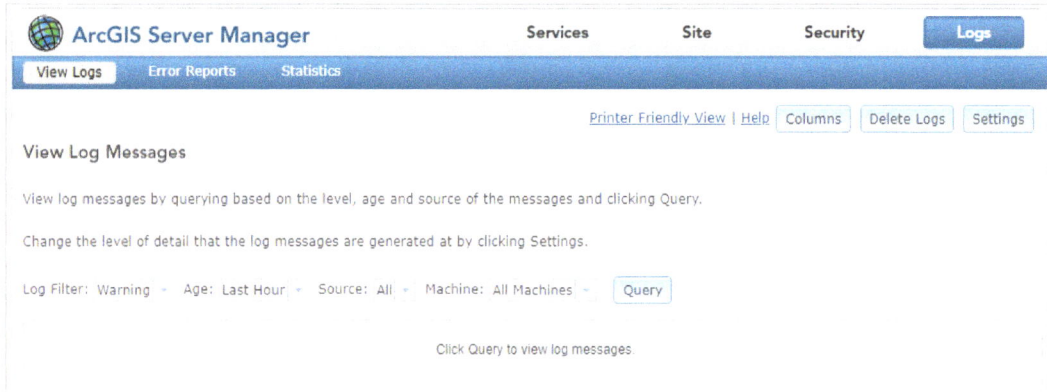

Figure 17.3. Logs for ArcGIS Server in ArcGIS Server Manager.

ArcGIS Server logs are accessed through ArcGIS Server Manager. These logs are helpful when troubleshooting access issues to databases, troubleshooting service failures, or unexpected geoprocessing service behavior. If ArcGIS Server Manager is inaccessible, these logs may be accessed in in the logs directory of the machine that hosts ArcGIS Server. These files may be viewed in any text editor.

ArcGIS Data Store

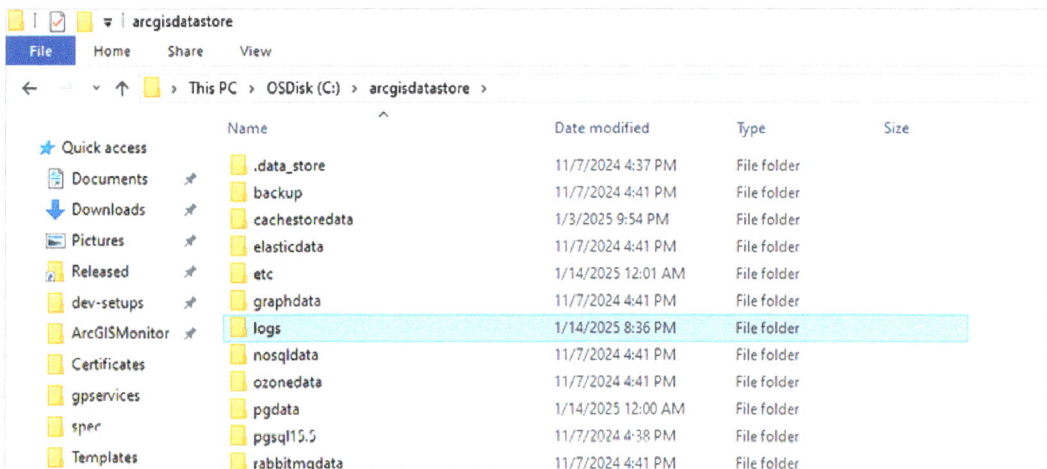

Figure 17.4. The logs that are captured by ArcGIS Data Store can be found in the machine in which the data store is stored.

ArcGIS Data Store is a bit interesting, in that its logs can be viewed in two locations. Because ArcGIS Data Store is used in the context of the hosting ArcGIS Server, most logs related to operating crucial features, such as hosted feature services and scene services, are displayed in ArcGIS Server Manager. However, ArcGIS Data Store also collects the same logs in its own logging directory within the ArcGIS Data Store machine. The best way to access this is through the installation directory on the ArcGIS Data Store machine and selecting the logging file that is the most relevant to your investigation time.

ArcGIS Data Store logs on the machine location collect information about the underlying tasks and processes required to make the data stores work properly. For example, if you are troubleshooting issues with a multinode deployment of the spatiotemporal big data store, it would be a good idea to compare each of the node's logging locations to learn more about what each of them is currently doing.

Logs are an important way to get an audit of system actions that are occurring as you reproduce the error. However, there is one more troubleshooting method that will serve to bring these logs together into a cohesive story.

Investigating the issue using browser developer tools

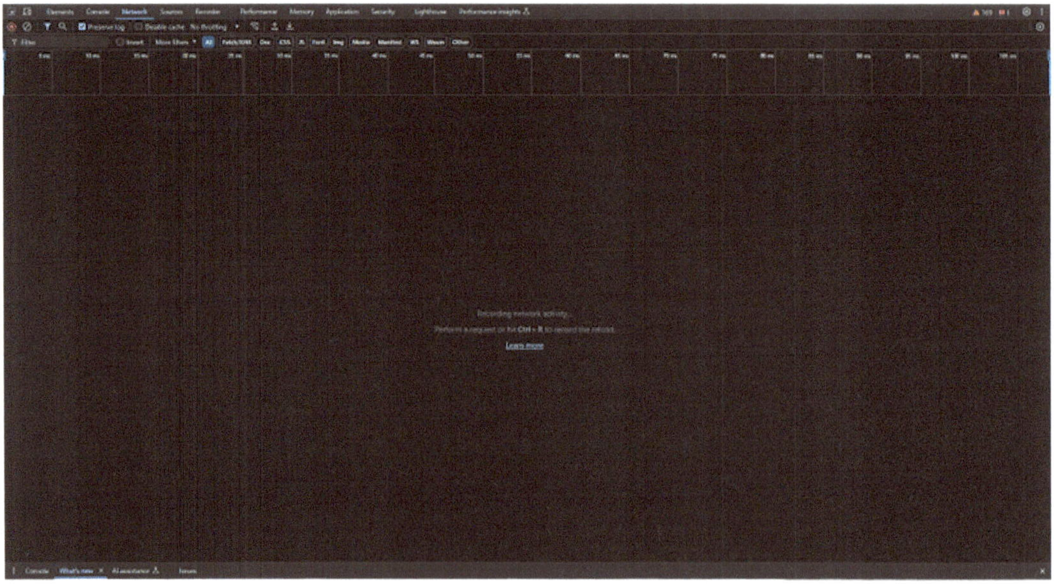

Figure 17.5. You can use tools, such as the developer tools in Google Chrome, to investigate issues while using web apps in ArcGIS Enterprise.

Browsers such as Google Chrome, Mozilla Firefox, and Microsoft Edge include built-in browser development tools that can be used to monitor web traffic from your browser session to different servers. Browser tools are an excellent way to immediately capture all relevant information about your session. These tools can include a set of important details that may indicate where the problem may lie. When you troubleshoot issues on the web apps in ArcGIS Enterprise, browser tools reveal the exact network calls that are breaking. In addition to broken calls, browser tools will also report on any loading errors appearing on the page. This is particularly helpful when troubleshooting widgets not loading correctly.

If working with Chrome or Edge, opening browser tools will reveal a steady stream of information about the network calls appearing on that page on the Network tab. Each entry is categorized—in most cases, the Name, Status, and Time categories are the most useful. Depending on the HTTP response of the request, you will see the request return in a different color. Successful, informative, and redirection responses will return in a green font, whereas failed or unauthorized status codes will return in a red font. At the top level, there are five HTTP response status code classes:

- Informational responses (100–199)
- Successful responses (200–299)
- Redirection messages (300–399)
- Client error responses (400–499)
- Server error responses (500–599)

Note: 200 requests will indicate that the server received a request and has completed and sent it back to the client successfully. However, administrators should inspect all 200 requests to verify that the operation itself was successful.

Clicking on any of these requests will open an additional window. In terms of information, this is where you will find the most details on the nature of the request and the response sent by the server. But what if you are investigating missing elements in a web page?

The **Console** tab reveals information related to specific warnings and errors that appear when different web page elements are loaded. In most circumstances, warnings and errors at this level are harmless; however, this is a good resource to check if different parts of a web page are not rendering properly or are missing.

Browser developer tools are a way to audit the network traffic that is occurring before, during, and after an issue happens. It allows the administrator to understand what leads to an issue, while also filtering out unnecessary background noise. Other developer tools have more functions, depending on the type of testing you're attempting; however, as a first response, browser developer tools are more than capable. Keep in mind that when you use

17

browser development tools, only traffic occurring between the client and the server will be captured, such as calls made from Map Viewer to ArcGIS Server for a feature service. These tools will not capture inter-component calls such as the types of calls the ArcGIS Enterprise portal makes to ArcGIS Server to authenticate someone accessing ArcGIS Server Manager. If this is required, consider using a stand-alone proxy tool.

Tying it all together

So far, we've discussed creating a first-response plan to gather details about unexpected behaviors in ArcGIS Enterprise, how to access and interrogate different logs, and how to use browser-based developer tools to glean more information about what is occurring on the network. Bringing these resources together to form a cohesive story about what happens in the environment requires a deep understanding of one element: time.

When you reproduce a problem with the intention of gathering network traffic and logs for later analysis, it is imperative to pay attention to the exact time that the test begins. Keeping track of the seconds and knowing the start times will save you a tremendous amount of time and effort in finding logs and tying them back to network calls.

Tens, sometimes hundreds of debug-level logs, may be generated per second if the system is under heavy load. Filtering these logs out by the time of interest will keep your analysis focused on specific actions and calls being made by your client when the issue occurs.

With the details above, a sample troubleshooting session might look like this:

1. A user discovers that a feature service is no longer accessible in Map Viewer.
2. An ArcGIS Enterprise administrator joins a conference call to observe and confirm the behavior.
3. The administrator reproduces the issue in their own environment.
4. The administrator sets the ArcGIS Server logs to Debug and begins capturing web traffic.
5. The web traffic reveals that an access error is hit when the service attempts to load.
6. Switching to ArcGIS Server logs, the administrator focuses on the time of the error to get information about which data source is causing problems.
7. Reviewing the data store, the administrator can confirm that the credentials used for access have been changed. Updating the credentials resolves the issue.

This chapter has presented an ideal scenario of how troubleshooting may lead to a solution for an access issue. Although this chapter is not an exhaustive guide on how to troubleshoot every issue that may occur within ArcGIS Enterprise, the steps outlined here are vital to establishing a solid start to resolving any technical issue. If you reach a point that an issue remains unclear or unresolved despite the investigation, you can turn to two additional resources:

- **Esri Community:** This free online platform connects users to ask questions, share ideas, and network. The knowledge base within the community is massive, encompassing a large scope of product communities, user communities, and regional user bases. The best way to approach researching with Esri Community is to search for an error message you've found on an appropriate product board. The ArcGIS Enterprise Product Community contains many subboards that touch on different technical areas. For example, if you are running into an issue with content management within the ArcGIS Enterprise portal, the ArcGIS Enterprise Portal Questions board will be the best place to start your search. It is also a good place to add new questions or get opinions on different workflows.

- **Esri Technical Support:** Esri maintains a department of staff worldwide that offers technical support on all products that are within an active life cycle. When working with support staff, users can expect to speak to a live person about their issue, screenshare if appropriate to investigate the problem deeper, and document and log defects if a bug is encountered in a workflow. If maintenance is active on your customer account, you may create a case by either calling in to the support line or creating a ticket through My Esri. Coming prepared to a support interaction with a clear problem statement, relevant logs, and system architecture diagram will yield a more efficient engagement.

Although troubleshooting in ArcGIS Enterprise may be challenging, a tremendous amount of resources are created every year to assist administrators in building their knowledge on this platform. Documentation updates, new ArcGIS blogs, technical papers, and new content in the ArcGIS Architecture Center are all created out of the growing need to address more advanced implementation of ArcGIS Enterprise. Staying current on these resources helps administrators stay updated on important updates in the technology, as well as different troubleshooting and monitoring methods.

17

Fictional user story

Medio County has seen an increasing rate of adoption of its publicly accessible Medio311 application. Medio311 lets citizens submit work orders to the county with picture attachments to a dedicated response team. Medio County recently experienced a polar vortex, which causes substantial damage to different parts of county infrastructure. Although emergency utility workers were able to secure and resolve most of the critical issues in the county, Medio311 began to receive a substantial increase in work orders.

On the second day of the weather event, Elise Medina, the county's GIS manager, learned that the application was no longer displaying some sublayers. Worse yet, one user reported that the application was no longer accepting new work orders.

Elise began her troubleshooting efforts with the work order submission issue because that was the most time-sensitive problem. She attempted to replicate the issue but was unable to submit test work orders as expected. Elise then called the impacted user to isolate the issue. The user was also unable to access Medio County's ArcGIS Enterprise organization, or any other web page. Elise helped the user determine that their issue was caused by a failure in their network connection, rather than with the Medio311 app, and connected them to a resource that could resolve that problem.

Moving to the failure of some sublayers to display, Elise replicated the issue in the Medio311 app. The same layers also failed to display in Map Viewer, which ruled out the Medio311 app itself as the source of the problem. Elise didn't publish these particular sublayers, so she wasn't familiar with their configuration. Using the developer tools in her browser, she saw the failed requests to the service REST end point. Navigating to that REST end point directly in her browser, Elise noticed that all the failing sublayers are part of the same service. That is a strong indication that a failure at the service level caused the problem. Checking the service configuration, Elise noticed that the service had stopped.

Elise knew that she didn't stop the service, and she couldn't think of a good reason anyone else would have stopped it either. Consulting the logs, Elise discovered warning messages indicating the service had stopped, which began around 11 p.m. the previous evening. The only other person that could have stopped the service is the manager of the Utility Department, Bob Gustafson, who is Elise's backup GIS administrator and also the owner of the service. Because she didn't know why Bob stopped the service, she contacted him first before turning it back on. During their conversation, Elise discovered that in the previous evening's scramble to deal with the impacts of the polar vortex, Bob had inadvertently turned off the service. Elise restarted the service, and the Medio311 app began working as expected.

Tutorial 17: Troubleshoot technical issues

Set an information banner

When you work through a major troubleshooting issue, communicating your response is vital toward maintaining transparency and accountability to your organization. An information banner is an easy way to acknowledge a problem to your user base, while also providing key follow-up information on what users should do while troubleshooting occurs.

1. Sign in to your ArcGIS Enterprise portal as an administrator.

2. On the navigation bar, click the **Organization** tab and then click the **Settings** tab.

3. On the left, go to the **Security** section.

4. Scroll down to the **Information banner** section and select **Set information banner**.

Information banner

Use information banners to alert your organization's members about your site's status and content, such as maintenance schedules, classified information alerts, and read-only modes. Banners will appear at the top and bottom of your pages.

Set information banner

5. Enable **Display information banner** to begin entering needed information. Configure an information banner that is appropriate for your organization.

> *Tip: Besides text, admins may also choose to change the background and text color of the information banner. The contrast ratio may also be changed to bring more or less emphasis to the banner.*

Work with logs

Next, you will learn how to access logging end points, change the logging level, and query logs.

1. Using an administrator account, access and sign in to the Portal Administrator Directory.

2. From **Resources**, select **Logs** from the list.

17

3. Select **Query**. At the bottom of the page, click the **Query** button.

4. On the resulting page, notice the default WARNING log messages that appear.
   • What are the different messages that you see?

5. Return to the **Logs** page and click **Settings**.

   This will present you with information on where the logs are being stored, the retention policy, and the logging level.
   • What is the current log level?
   • Where are the log files stored?

6. Click **Edit** and set the **Log level** to DEBUG.

7. Navigate to the **Query** page and ensure that the **Log Level** setting shows **Debug**.

   Note: Although logs may be collected at any level, it is highly recommended that the DEBUG log level is only used during active troubleshooting, ideally after hours.

8. Query the **Debug** level logs.
   • What type of information do you see displayed?

9. Navigate to the **Settings** and reset the **Log level** to the previous setting.

   The ArcGIS Server Logs are in ArcGIS Server Manager.

10. Navigate to the hosting server manager for your ArcGIS Enterprise organization.

11. On the top, click the **Logs** tab.

    As in the ArcGIS Enterprise portal, the ArcGIS Server Logs are set to the **Warning** logging level by default.

12. Click **Settings** and change the **Log level** to **Debug**.

    Querying the logs at the **Debug** logging level will reveal all logs captured at **Debug** or higher.

13. Click **Settings** again to set the logs back to the previous setting.

Check the sync status of distributed collaborations

If your ArcGIS Enterprise organization has a distributed collaboration set up with another ArcGIS Enterprise organization or an ArcGIS Online organization, distributed collaboration sync status logs may provide key insight into identifying and troubleshooting sync issues.

1. Navigate to the ArcGIS Portal Directory.

2. In the top-right corner, click **Login** and sign into the ArcGIS Portal Directory.

> *Tip: For example, if your Service URL is https://gis.example.com/server, you can access the ArcGIS Portal Directory at https://gis.example.com/portal/sharing.rest.*

This page contains the basic user level information for the administrator account.

3. From the **User** information, click the **OrgID: 0123456789ABCDEF**.

4. Scroll down to the bottom of this page. From the **Child Resources** list, select **Collaborations**.

This screen will contain information on active distributed collaborations within your organization. Clicking on different collaboration IDs will display an option to select logs.

5. If a collaboration is present, click the **Collaboration ID**.

6. Click **Collaboration Workspaces**.

7. Click **Collaboration Workspace ID** and select **Sync Status**.

The **Sync Status** logs will reveal information on a set of tasks that are required for a distributed collaboration to work. These tasks include the replication package import/export status, export package replication, and reachability of the guest/host peer.
- What other items do you see in these logs?

17

Configuration reporter for ArcGIS Enterprise

The configuration reporter is a Python script that was developed by Esri to aid in the troubleshooting of environment-specific issues. When run on supported versions of ArcGIS Enterprise, the configuration reporter reports on logical architecture of the deployment, licensing and certificate information, and the configuration details of ArcGIS Server and Portal for ArcGIS components.

1. Before beginning, read the "Use the Configuration Reporter" article, located at links.esri.com/GTKEnterprise-ConfigReporter.

    This article includes important information relating to the requirements to run the tool, pre-requisites including Python, and ArcGIS Enterprise version requirements.

2. From the article, download the **ConfigReport.zip** file.

3. Extract the **ConfigReport.zip** file and open the **ConfigReport** folder.

    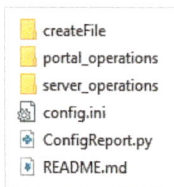

4. Open the **config.ini** file in a text editor (Notepad or Notepad++ is recommended).

5. Add your **target_portal** URL (portal web adaptor URL), your administrator username, and password.

6. Enter the correct **validateCertificate** value:
    -   If the certificate is trusted, enter the value **True**.
    -   If the certificate is invalid or being accessed via the IP or host name instead of the name on the certificate, enter the value **False**.

7. Save the file and close it.

8. Open the Command Prompt as an administrator and change the directory to the location of the **ConfigReporter**.

9. Run the script, paying close attention to use the ArcGIS Server or ArcGIS Pro Python environment.

   The output of the script is in arcgis_reporter-main > ConfigReport > generated_reports.

10. Navigate to the **generated_reports** folder and check the contents.
    • What information do you see with respect to your deployment?

Summary

This chapter covered the basics on how to respond to a technical issue in your ArcGIS Enterprise organization. By observing the issue, you gained an understanding of the nature and reach of the behavior. Isolating the problem to a specific service or data store gave you key insights into what may be the root cause of the problem. Categorizing severity informed the level of effort you and your IT team may need to resolve the issue. And finally, communicating the problem to key stakeholders set up awareness and accountability within your organization.

We finished this chapter by reviewing the logging capability built into ArcGIS Enterprise, as well as additional tools that are commonly used to understand the underlying network traffic that is at play when actively troubleshooting a problem. Knowing where to look and analyzing the patterns found between logs may lead to a solution. However, Esri Community and Esri Technical Support are two resources to consider when working through a troubleshooting scenario.

17

CHAPTER 18

Backing up and restoring ArcGIS Enterprise

Objectives

- Define what a backup strategy is in context of ArcGIS Enterprise.
- Understand the differences between content-, system-, and data-level backups.
- Examine how to orchestrate a backup strategy.

Introduction

Complex enterprise systems rely on a wide variety of interlaced applications, data, and dependencies to serve information to end users. If something goes wrong in day-to-day operations, it's imperative for administrators to create a vigorous business continuity and disaster recovery (BCDR) strategy. Esri recommends that ArcGIS Enterprise administrators have backups before upgrading ArcGIS Enterprise, installing software or system patches, applying any major network updates to the system, and as critical hosted data changes. A well-documented and verified backup and disaster recovery plan is vital to the long-term success of any ArcGIS Enterprise deployment.

This chapter will outline the basics of what a comprehensive backup strategy may look like in an ArcGIS Enterprise deployment. We will define the goal of the backup process and what kinds of tools exist to achieve system recovery. We will introduce the Web GIS Disaster Recovery tool (WebGISDR), a tool provided by Portal for ArcGIS that creates organization backups of ArcGIS-managed data. This chapter will provide an overview of how other backup activities complement WebGISDR to achieve content consistency across your organization.

By the end of this chapter, you will be able to articulate the goals of a backup process and begin to have conversations with stakeholders to achieve and verify a well-documented BCDR process that is scoped properly and reproducible. Through the application of this plan, administrators can expect to save on downtime and overall effort when system outages or data loss require intervention.

Why should you create a BCDR strategy?

The ArcGIS Architecture Center states, "For enterprise systems with availability expectations, requirements, or commitments, a clearly defined, actionable, and well-tested backup and disaster recovery (BCDR) approach is critical." Backing up content, projects, and data connections prevents inadvertent data loss and project delays. In the best-case scenario, losing these items leads to wasted time and lost revenue and rebuilding the system from scratch. The worst-case scenario may pose real-life consequences for those who rely on these systems during crisis. A well-practiced and verified backup strategy provides your user base with peace of mind that a contingency plan is in place when the unexpected occurs.

The contingency plan for a BCDR scenario will consider three questions:
- **Scope**: Which systems will be included in the backup process?
- **Method**: How will the backup process be technically achieved and maintained?
- **Location**: Where will the backup be stored?

Each of these pillars must be considered within the context of your implementation of ArcGIS Enterprise and weighed against the overall cost of creating and maintaining a backup for each system. For example, a fleshed-out backup strategy will usually be applied to the production environment, including ArcGIS Enterprise, underlying databases and file stores, as well as machine images. Answering these questions represents the scope of the backup process, so let's take a closer look at what this means.

Backup scope

To address the scope of a backup, ArcGIS Enterprise administrators must consider the sources of the data that exists within ArcGIS Enterprise, the systems that host ArcGIS Enterprise, and ArcGIS Enterprise as a system. Additionally, if tiered environments (such as development, quality assurance, and production) are used to promote content to a public-facing deployment, each tier should have a backup set up as well. It is helpful to imagine the scope of a backup in three categories: data-level backups, application-level backups, and system-level backups.

ArcGIS Enterprise exists within the context of an organization's network, which includes other enterprise-grade pieces of software. One of the main purposes of using ArcGIS Enterprise is the ability to create content from a variety of data sources, such as databases, file shares, and object stores. As we covered in chapter 10, referenced data coming from these data sources is considered user-managed data. This includes creating and managing backups of relevant databases, referenced folders, and caches. Within the scope of a backup strategy, these are data-level backups.

For ArcGIS-managed data such as hosted feature services, ArcGIS Enterprise portal content, and published geoprocessing tools, administrators can rely on the built-in WebGISDR utility. The WebGISDR will create a content and configuration backup of an ArcGIS Enterprise deployment, which may be used to restore a system in the event of an error. Hosted feature layers, apps, web maps, groups and users, services, webhooks, custom data feed (CDF) providers, server object interceptors (SOIs), and server object extensions (SOEs) are included in a WebGISDR backup file. The WebGISDR workflow represents an application-level backup.

All ArcGIS Enterprise components rely on systems or hardware to run properly. This includes data stored on disk as well as memory. Depending on the underlying infrastructure, these systems may be backed up through snapshots or machine images. Although the ways to complete this may vary, system-level backups are important to consider when scoping out a backup plan, because it could drastically reduce downtime when attempting recovery.

The combination of data, application, and system level presents the total scope of a backup. Depending on your system usage and the capabilities being employed by ArcGIS Enterprise, you may not need all levels of the backup to reach your recovery objectives. For example, let's say that you have configured ArcGIS Server with a raster processing role within ArcGIS Enterprise. This site does not include any of its own data, user managed or otherwise. As such, a system-level backup is enough to preserve the function of the server and reduce downtime by excluding unnecessary restore actions. This illustrates the importance of understanding what each component of ArcGIS Enterprise is responsible for and what needs to be backed up for the system to function. Once a scope is established, administrators must create and document methods.

Backup methods

The backup methods that you choose should be informed by your organization's needs when using those backups to recover your systems. Recovery Point Objective (RPO) is a measure of how much lost work you consider acceptable when recovering from backups. Any work completed since the last backup will be lost. The less work you can afford to lose, the more frequently you must create backups. Recovery Time Objective (RTO) is a measure of how much system downtime you consider acceptable while you restore from a backup. Restoration is not an instantaneous process, and some strategies take longer than others. The right RPO and RTO will vary by organization, but the optimal values are unlikely to be zero. Ensuring lower RPO and RTO values requires greater financial outlays, higher system complexity, and more staff time, so you will want to find the right balance between those needs.

18

Backing up ArcGIS Enterprise and other systems identified in the scope stage requires careful coordination and synchronization. A referenced feature service exists as an item within the ArcGIS Enterprise portal, a service on ArcGIS Server, and a table within an enterprise geodatabase. This means that for a backup to be content consistent, backup activities need to occur when the system is not experiencing any editing activities. This means that application, data, and system- level backups should be taken at roughly the same time, when the organization is not in a state of active editing. Synchronizing your application, data, and system-level backups is essential to ensure that no data loss occurs. Part of ensuring synchronization is understanding the activity occurring on ArcGIS Enterprise. For example, if new services aren't published and service configuration is static, only the data tier is important to back up or restore, depending on the circumstance of recovery.

Note: When you run any backup task, ensure that the machines are not under significant load, and that all change activity has been suspended. Certain backups may require that the services running ArcGIS Enterprise have been turned off to ensure that no changes are being made to the environment during a machine snapshot process.

The methods each administrator employs for their ArcGIS Enterprise organization will vary, depending on your underlying systems and data storage choice; however, let's review a high-level overview of what these choices may look like.

Application-level backups: We introduced the Web GIS Disaster Recovery tool (also known as WebGISDR) earlier in this chapter as ArcGIS Enterprise's method of creating an application-level backup of its hosted content and configuration. The WebGISDR works by launching component-level, synchronous backup jobs across an ArcGIS Enterprise environment. By design, this means application-level backups completed with the WebGISDR tool achieve the highest level of content consistency. The utility will take the output of backup processes running on ArcGIS Server, Portal for ArcGIS, and ArcGIS Data Store and bundle them in a single file, which can be used to restore content, users, and different configurations later. The entire machine's file system will not be backed up by the WebGISDR utility.

The WebGISDR utility has three methods of creating a backup of ArcGIS Enterprise: full, incremental, or backup. Running the tool using the full backup option will create a complete backup of ArcGIS Enterprise. If running the utility for the first time, you must create a backup using the full or backup modes.

Running the WebGISDR in the incremental mode will consider changes made only after the latest full backup of ArcGIS Enterprise. Incremental backups take less resources and less time, enabling ArcGIS Enterprise administrators to create more frequent backups. The only drawback to applying incremental backups comes in the form of restoration time: If recovering from a disaster, administrators may need to perform the restore operation twice—first to apply a full backup and then to apply the subsequent incremental backup.

To implement a common application of the WebGISDR workflow to balance creating backups with system performance, some ArcGIS Enterprise administrators will opt for a combination of full and intermittent backups. On off days when system usage is low, such as the weekend, they may run the WebGISDR in the full mode to keep an up-to-date full background in store. During the weekdays, running the WebGISDR utility in the intermittent mode will keep the full backup current and avoid a larger resource cost.

Note: ArcGIS Enterprise on Kubernetes has a different backup process that may be run from Enterprise Manager. This backup would be the equivalent of a full backup; intermittent backups are not currently possible. More details on this workflow can be read in documentation titled "Data Loss and Downtime minimization."

Data-level backups: These backups can refer to file stores, cloud storage, and databases. When scoping and applying methods to ensure data consistency, an ArcGIS administrator should understand where critical data is stored. Collaborating with IT teams as well as database administrators on securing your key referenced and user-managed data stores is important to ensure complete data consistency across your organization.

File shares and folder structure backups depend on the underlying system servicing those stores. For example, network accessible storage (NAS) devices and storage area network (SAN) volumes must be backed up. An ArcGIS Enterprise administrator's responsibility in this context is to identify file shares and communicate with their IT team to strategize backup scenarios that work in tandem with their comprehensive BCDR plan.

Supported databases, including Microsoft's SQL Server, Oracle Databases, and PostgreSQL, have their own backup processes and procedures. ArcGIS Enterprise administrators, as they would with file shares, must identify geodatabases that are linked to their GIS services with the database administrator to ensure backups have been created for them.

System-level backups: These backups cover the underlying infrastructure that runs ArcGIS Enterprise, as well as other critical components. This can cover the virtual and physical hardware and file structures. Because ArcGIS Enterprise is capable of being deployed on many different systems, creating system-level backups must be coordinated through careful collaboration with your IT department. However, there are some best practices to keep in mind when creating system backups.

18

> *Tip: System snapshots are an easy way to create a point-in-time recovery that can be applied quickly and efficiently to a system. System snapshots rely on the overarching virtual machine hypervisor to remain functional to keep working properly. In certain scenarios where a true disaster recovery is possible, IT teams need to plan diverse backup options to complement snapshots to recover from these scenarios.*

Two factors must be considered when taking a system-level backup. The first is crash consistency. ArcGIS Enterprise documentation states that if a system backup is taken of a machine running ArcGIS Enterprise, the underlying services must be in a stopped state. This prevents system files from being captured while they are being edited. Restoring to a system backup where the services running ArcGIS Enterprise were still active may render the system unusable. When a system-level backup is taken, this strategy represents a crash-consistent backup.

The second factor to consider is application consistency. ArcGIS Enterprise administrators commonly pair full WebGISDR workflows with a system backup. This ensures content available in ArcGIS Enterprise at the time of the backup will have the necessary system files captured by the machine snapshot process. A sample of this workflow may look like this:

1. Communicate system downtime and set ArcGIS Enterprise to read-only mode.
2. Take a full WebGISDR backup of ArcGIS Enterprise.
3. After the WebGISDR is completed, stop all ArcGIS Enterprise services.
4. Take snapshots of all machines participating in the ArcGIS Enterprise organization.
5. Turn services back on.
6. Patch your machines (depending on the time passed).
   a. If patches fail, restore to snapshot.
   b. If the underlying systems fail, restore to WebGISDR.
7. Verify system health.

System backups and snapshots are an important part of creating a holistic backup strategy for any production ArcGIS Enterprise environment. Creating and documenting methods with application-level backups and data-level backups will enable administrators to save critical workflows in the event of a disaster.

Most of the backup methods can be automated. The benefits of automation include consistent timing of backup tasks, less administrative overhead, and establishing a set of backups that can be deployed to if problems continue to exist within the latest backup. An example of this is using Windows Task Scheduler to automate the WebGISDR utility at the end of the business day. The last part of a well-made backup strategy is choosing the location of your backups.

Backup location

An often-overlooked part of the backup process is an isolated and documented location of backup files. Although storing the output of the WebGISDR on the host machine of ArcGIS Enterprise may be more convenient and more streamlined, there is a risk of losing this file if the underlying machi0ne that powers ArcGIS Enterprise fails, or if the environment falls prey to a ransomware attack. ArcGIS Enterprise administrators should consider different storage systems to store the different backup types discussed earlier to add a layer of redundancy to application, data, and system backups.

Restoring after a backup

So far, we've covered why administrators should create sound backup strategies. We also talked about how to scope out different levels of backups and what ways and means may be applied to reach content, system, and data consistency. Now that we've covered how to create a backup, let's discuss the fundamentals of creating a Disaster Recovery (DR) plan and restoring after a backup.

The ArcGIS Architecture Center draws a strong relationship between a DR process and a backup-and-restore operation. Detailed terminology and concepts on DR workflows may be found in the backups and disaster recovery section of the ArcGIS Architecture Center; however, the basic principle lies in creating a DR process that balances a complete backup of the content with minimal downtime. Before a disaster strikes, administrators should document the location of their backups in the same order that the systems will be brought back online.

In a recovery scenario, not all systems may have been affected by the outage. During the incident, identifying which systems were affected may require only certain recovery tasks. Let's look at a set of scenarios:

- **Accidental critical hosted feature layer deletion:** While cleaning up an organization's redundant hosted feature layers, an administrator accidentally deleted a critical hosted feature layer. The administrator is able to apply a WebGISDR backup to recover the recently deleted hosted feature layer.
- **Problematic system update:** While installing a service patch for the system running a production ArcGIS Enterprise deployment, an administrator notices that the underlying services are no longer turned on. The administrator can work with their IT team to restore to a recent machine snapshot to regain system function.
- **Unexpected system failure:** A natural disaster caused a power surge that knocked out different machines within a data center. As the backups were kept offsite, the IT team, database administrator, and ArcGIS Enterprise administrator restored access to the machines from offsite backups. Referenced datastores were restored first, followed by the system running ArcGIS Enterprise, and completed with the WebGISDR system restore.

18

Each of these scenarios serves to outline the importance of understanding when restoring a system is applicable. Additionally, the third scenario outlines an important order of operation when you conduct a DR: All underlying system dependencies, such as file shares and databases, need to be restored before restoring ArcGIS Enterprise. Once the restore is complete, administrators should verify the function of their organization by working with content owners and publishers to confirm the function of their projects and services.

Fictional user story

Medio County recently underwent a disaster recovery readiness audit. The result of this audit revealed that their current backup strategy for ArcGIS Enterprise was not running frequently enough to keep up with the many changes and edits being made. Elise Medina, the GIS manager, created the initial backup strategy when ArcGIS Enterprise was used in production. This backup plan included a single WebGISDR backup per week alongside a machine image of all the machines participating in the base deployment of ArcGIS Enterprise.

As Medio County's implementation of ArcGIS Enterprise grew, so did the scale and scope of the environment. Elise found that there were three referenced datastores being used in the environment, which did not have clear backup policies in place. Additionally, she found that her default location for storing the WebGISDR utility outputs was on the same machine that was running the ArcGIS Hosting Server. This represented a possible failure point if that machine became inaccessible.

With this knowledge, Elise approached Jake Nilsen, the IT director of Medio County. She began by explaining the current state of the backup and how the scope has changed over time. She articulated the importance of backing up the referenced data stores and that Jake would need to take an active role in coordinating database administrators to take action. She finished her presentation by providing Jake with an architecture diagram of the deployment, outlining the critical systems that did not currently have adequate backups.

Over the next few weeks, Jake and Elise worked together to create and document methods that would allow them to recover from a disaster in short order. Jake worked with different stakeholders in the IT department to establish regular content backups for their databases and a system based on a redundant array of independent disks (RAID) for their critical data stores. Elise worked to refine the WebGISDR process by storing the backup off the machines running ArcGIS Enterprise. She also automated the utility to take daily intermittent backups in addition to a full backup each weekend. With the help of Elise, Jake also created an automated task that would take a machine snapshot of each machine running ArcGIS Enterprise. Through their work, Jake and Elise substantially improved their readiness to deal with an unexpected "production down" scenario by automating more frequent backups and creating a more holistic backup strategy.

Disaster recovery can be a stressful experience and include many moving parts, iterations, and time sensitivity. A successful disaster recovery aims to minimize system downtime while recovering as much business-critical content as possible. Throughout the process of scoping out the system backup and creating backup methods, documenting the plan is a key piece in speeding up the disaster recovery process. Besides documenting the workflow, practicing and verifying backup methods will further improve the confidence of the DR process.

Tutorial 18: Investigate ArcGIS Enterprise settings for business continuity and disaster recovery

In this exercise, you will confirm how your ArcGIS Enterprise components are configured to enable your organization's BCDR plan.

ArcGIS Data Store

1. Sign in as a user with administrator privileges to the machine where the ArcGIS-managed relational data store is located.

2. Open a command line shell (such as cmd or PowerShell on Windows or bash on Linux) as an administrator and navigate to the directory where ArcGIS Data Store is installed.

   | Tip: See chapters 2 and 3 for information on installation directory locations.

3. Navigate to the **tools** directory and list the contents of that directory.

   The scripts in that directory provide the management capabilities for ArcGIS Data Store.

18

4. Execute the **describedatastore** script.
   - In addition to the relational data store, are there any other ArcGIS Data Stores installed on this machine?
   - For each data store, is there a backup location and schedule configured?
   - For each data store with a backup location specified, where is that location physically located? You may need to work with your IT administrators to confirm the location.
   - For each data store with a backup schedule configured, how frequently are backups taken and how long are those backups retained?
   - Is point-in-time recovery enabled for any of the data stores?

5. Repeat the steps above for any ArcGIS Data Stores installed on other machines.
   - Given your organization's recovery point and recovery time objectives, are these settings adequate to meet your needs?

WebGISDR

1. Log in as a user with administrator privileges to the machine where Portal for ArcGIS is installed.

2. Navigate to the directory where Portal for ArcGIS is installed and then navigate to the **tools/webgisdr** directory.

3. Open the **webgisdr.log** file and look for the entry for the most recent successful EXPORT.
   - When was the last time WebGISDR was successfully run?
   - Where is the backup location?
   - What kind of operation was run (backup, full, or incremental)?
   - How long did the backup take?

4. If at least one backup has been made, navigate to the backup location.
   - Based on the dates of backups, are backups being taken on a regular basis?
   - Given your organization's recovery point and recovery time objectives, is WebGISDR being run in a way that is adequate to meet your needs?

Components not covered by WebGISDR

The WebGISDR utility does not back up every component used by ArcGIS Enterprise. You will need to use other tools to monitor backups for these components.

1. If you have any map service cache tiles, hosted tile layer caches, or file-based user-managed referenced data sources, confirm the location of any backups for the directories that contain these files.

2. For user-managed referenced databases, confirm the backup settings of your database management system.

3. For the ArcGIS-managed spatiotemporal big data store and graph store, use the **describe-datastore** utility to check backup settings.

4. For federated ArcGIS Mission Server, ArcGIS Notebooks Server, or ArcGIS GeoEvent Server sites, consult the documentation for those server roles.

Snapshots

If your organization uses virtual machine snapshots as part of your BCDR plan, work with your IT department to confirm the scope of those backups, where the snapshots are stored, and whether the snapshot configuration is adequate for your needs.

Summary

In this chapter, we identified how to build a holistic backup strategy for an ArcGIS Enterprise organization. We covered the three main domains of a backup, including the scope of the backup, methodologies in applying a backup, and how to choose the location for backup media. Additionally, we discussed three technical areas where backups may be necessary in ArcGIS Enterprise: application-, system-, and data-level backups. We concluded by covering how to restore from your set of backups and how to identify the components of ArcGIS Enterprise that may need to be backed up.

18

CHAPTER 19

Patching ArcGIS Enterprise

Objectives

- Learn about patching principles for connected and disconnected environments.
- Understand how to prepare ArcGIS Enterprise before installing a patch or an update.
- Apply an available patch to your ArcGIS Enterprise deployment.

Introduction

Esri is committed to providing users with the most secure and issue-free experience across the ArcGIS platform. Before a new version of ArcGIS Enterprise is released, the software is tested through a set of internal quality assurance and quality control (QA/QC) tests and processes to ensure quality. At times, beta testing and other supporting events occur months before release to stress test implemented defect fixes and new features in real-world scenarios. Before the latest version of software releases, Esri completes a thorough certification process to ensure that the software will reliably run on all supported systems.

Esri's commitment to software development does not stop at the release of the latest version of ArcGIS Enterprise. New defects and security issues are certain in an actively changing technological landscape. In response, patches and updates are vital to maintain the security and function of all software. The ArcGIS Enterprise life cycle defines all versions of ArcGIS Enterprise in general availability and extended support to receive software patches and updates.

Patching for Windows and Linux variations on ArcGIS Enterprise is delivered as executables that can be run on the machines that host the installable software components of ArcGIS Enterprise. Because the product life cycle for ArcGIS Enterprise on Kubernetes is more accelerated, patches are delivered through system updates in ArcGIS Enterprise Manager. This chapter will focus primarily on applying Windows and Linux patches; however, communication principles will still be relevant to ArcGIS Enterprise on Kubernetes.

By the end of this chapter, you will know how to get information about new patches for ArcGIS Enterprise, as well as how to install these patches properly. Additionally, we will present an ideal patch installation workflow that will help administrators be accountable to users of ArcGIS Enterprise.

Patching basics

Part of the core mission of all software providers is to create and maintain a secure and issue-free environment for their user base. They do this through regular patch releases that resolve critical software defects and security vulnerabilities. Patches can and should be applied to every part of an enterprise-grade system as they become available. For example, if Esri releases a security patch, this should be applied within one month of release to maintain a secure environment. Similarly, if the operating system provider releases a security patch, the same rule applies. Although the source will vary, developing a consistent plan for applying patches is crucial to establishing good habits and recovering from unforeseen circumstances.

Patches developed for ArcGIS Enterprise by Esri typically fall into one of two categories: security patches and feature patches. Security patches are regular updates that address security-related problems within ArcGIS Enterprise. Feature patches include resolutions to issues that affect the usage of an application or component within ArcGIS Enterprise. Although these patches are named for a function or feature that may not be actively used within your workflow, it is best practice to install these patches as well. Most times, these patches will include important updates that are closely related to the feature and will improve the overall functionality of ArcGIS Enterprise. Because it is considered best practice to apply patches to ArcGIS Enterprise once they are available, it is essential to learn when patches have been released.

> *Tip: Some patches are cumulative, meaning that they will contain fixes presented in previously released patches. This can be verified by referring to the "Issues Addressed with this Patch" section of the description page for patches on Esri's technical support site. Because patches are cumulative, older versions or redundant patches are marked as obsolete.*

If your system allows access to https://*esri.com/ from the machines running ArcGIS Enterprise, you can use the ArcGIS Enterprise Patch Notification utility.

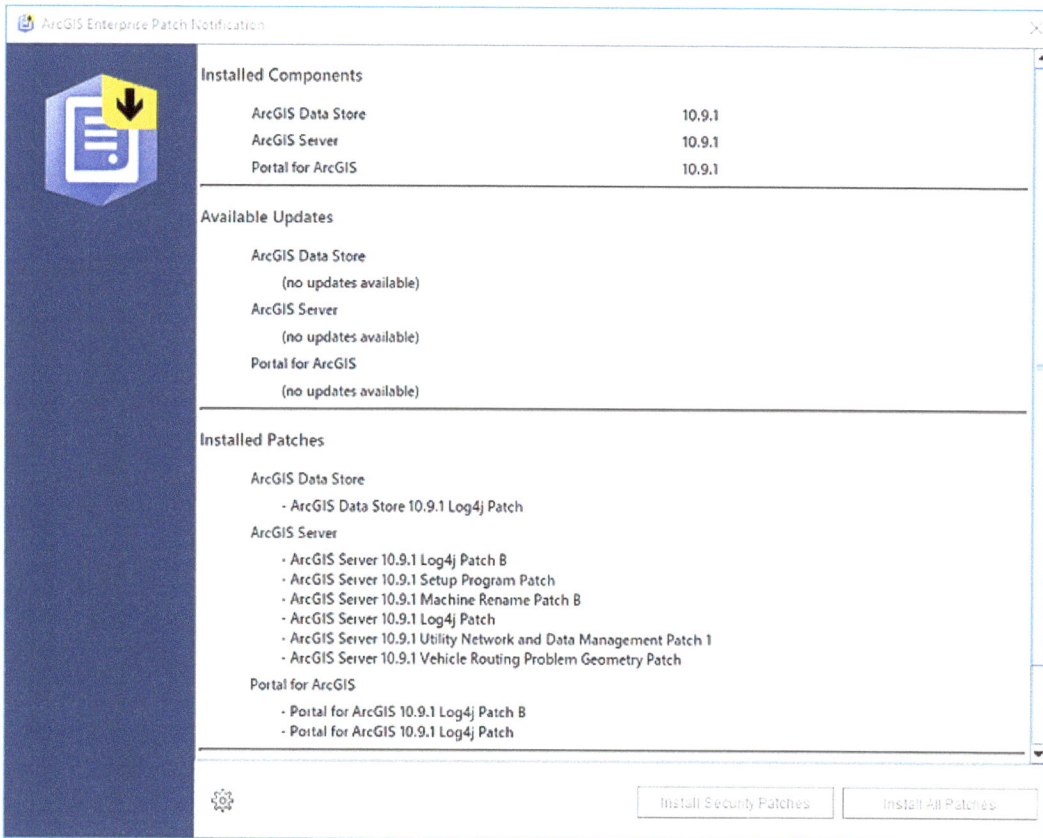

Figure 19.1. The ArcGIS Enterprise Patch Notification utility allows you to quickly install any patches available for your ArcGIS Enterprise deployment.

The ArcGIS Enterprise Patch Notification utility displays the version of the ArcGIS Enterprise component, available patches for installation, and installed patches. Each installed component of ArcGIS Enterprise comes with an instance of the ArcGIS Enterprise Patch Notification utility, so if you install ArcGIS Server and ArcGIS Data Store, you should expect to see two listings for the utility. The patch notification utility is the easiest way to get patch availability information. However, what if you want to find out more details on available patches, and the ArcGIS Enterprise organization is not able to reach out to any domains outside the network?

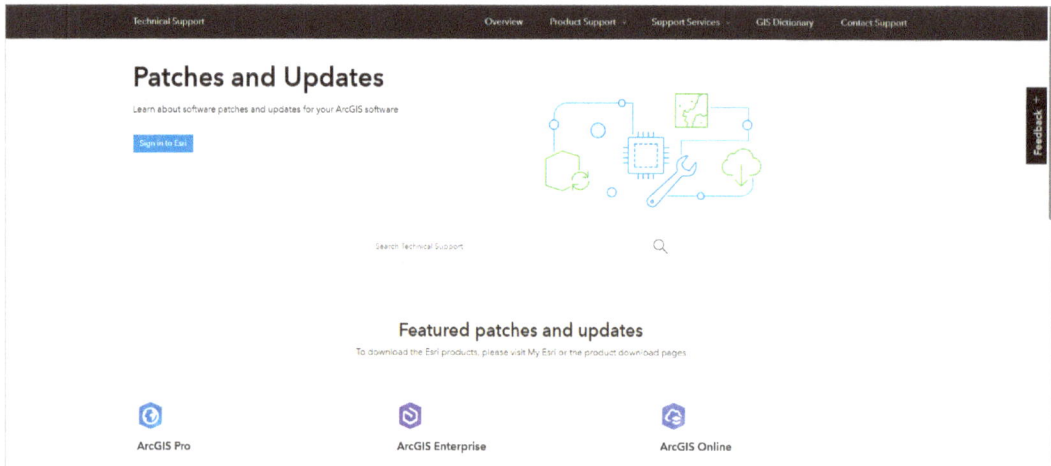

Figure 19.2. This Esri Technical Support web page highlights all patches and updates for ArcGIS products.

The Esri Technical Support web page for patches and updates at support.esri.com/en-us/patches-updates is the primary source of all patches released for ArcGIS Enterprise. Users can filter the site according to the ArcGIS Enterprise component and version. For notification purposes, you may choose to receive emails from Esri related to new patches being released in My Esri, a self-service portal for accessing Esri products and services. Each patch that is released has a corresponding web page that includes information about the name and version of the patch, release date, summary, issues addressed list, and detailed instructions on how to download and install the patch. Installing patches on ArcGIS Enterprise will incur system downtime. The next section of this chapter will review the steps administrators can take to ensure that patching is successful.

The ArcGIS Trust Center will also release trust advisories for major security patches on its page at https://trust.arcgis.com/en. Configuring an RSS feed to forward live updates on this page will make monitoring easier for security-minded administrators.

> Tip: In addition to including new development features, newer releases of ArcGIS Enterprise will include all security patches without needing to go through a patching cycle. This means that system uptime may be achieved quicker if you install the latest version of ArcGIS Enterprise, as opposed to deploying an older version that may have a larger patch footprint.

Considerations before patching ArcGIS Enterprise

Before installing any patches on either the underlying system or to ArcGIS Enterprise, administrators should consider and take the following precautionary steps:

1. **Back up ArcGIS Enterprise:** Verifying that ArcGIS Enterprise has been backed up within an acceptable time frame to your user community can save time and effort in recovery if any anomalies are encountered during patch installation. If your user community is particularly active, it may be necessary to set the organization to read-only mode before installing the patch for content consistency. For more information on establishing sound backup practices, see chapter 18.

2. **Test patch installation in lower-tier environments:** It is best practice to apply any major software and system changes to nonproduction environments to test the installation process and understand how long system downtime will last. Additionally, it's an opportunity to check on critical projects and workflows to verify their function before moving on to production. ArcGIS Enterprise administrators should apply ArcGIS Enterprise patches to development or quality assurance environments to understand the estimated downtime of the system during patch installation, verify the patch resolved the expected errors, and catch and address any irregularities before committing to production. At this stage, it is good practice to conduct the confidence tests you run on the system when changes, such as the application of patches, occur.

3. **Communicate and coordinate downtime**: Depending on the version of ArcGIS Enterprise, installing patches may take a substantial amount of downtime. Enhancements have greatly improved the efficiency of patch installation to all components in ArcGIS Enterprise. However, downtime should still be expected and communicated. Administrators should communicate potential downtime to their user base with informational banners and any other internal communication channels. Having an IT member available to address any added challenges when applying patches to a production environment is a good preventive step.

Following these steps before applying any patches to ArcGIS Enterprise will yield a solid backup point, a general idea on how long a patch may take to install, and transparency with your user base on how long ArcGIS Enterprise may be unavailable when applying the patch. Once this state of readiness is achieved, patches can be applied safely to a product deployment.

19

Applying patches to an ArcGIS Enterprise component

Installing patches requires access to the machine that has either ArcGIS Server, Portal for ArcGIS, ArcGIS Data Store, or ArcGIS Web Adaptor (IIS). The Patch and Update page for the selected page may identify additional prerequisites before applying a patch. As an added level of security, each patch contains a checksum that can be used to verify that the downloaded patch matches what Esri has released.

Each patch may apply to multiple versions of ArcGIS Enterprise on either Windows or Linux. Once the proper file is verified and placed on the target machine, it can be installed. There are differences to the Windows and Linux user experience; it is best to consult the patch documentation for specific steps. Once the patch installs, the service running the ArcGIS Enterprise component will restart, completing the installation.

If you are applying patches to multiple components at once, consider these additional factors when installing patches:

- **Single and multiple machine ArcGIS Enterprise Environments**: Patches may be installed on any components in any order.
- **Multiple machine ArcGIS Server sites**: Ensure that one machine is always available; patches may be applied to all other machines in any sequence.
- **Highly Available (HA) ArcGIS Enterprise Environments**: Administrators may choose to promote a standby ArcGIS Enterprise portal machine during the upgrade process of the primary to ensure uptime during patch installation. If this is not possible, the standby should be updated first. The same applies to the hosting ArcGIS Data Store machines, although ArcGIS Server may be updated in any order, as long as at least one machine is active.

Once patches have been installed, administrators should test ArcGIS Enterprise's basic functions. Publishing hosted feature layers and editing referenced feature services will verify whether all components in ArcGIS Enterprise are functioning as expected and have retained the access permissions they need to function. Looking through ArcGIS Server and ArcGIS Enterprise portal logs for any outstanding or unexpected errors can verify the absence of any outstanding errors.

Patching is an important maintenance task that should be completed on a regular basis. IT teams and administrators should work together to create reasonable opportunities to apply patches to your system. Documenting and standardizing the patching process will ensure a repeatable experience that may avoid unnecessary downtime and limit potential failures.

Fictional user story

The SuperBiz ArcGIS Enterprise organization has an international user base of more than 5,000 users. With a high level of adoption, system uptime is a vital goal for the long-term success of the ongoing projects within the system. Viktor Dudnyk, the ArcGIS administrator, and the IT team have deployed a two-tiered environment that includes a development and production environment.

Over the past week, Esri released a Portal for ArcGIS security update patch. This patch addresses a set of defects that are related to the overall security of the environment, as well as other supportive defect resolutions. Viktor received an advisory note from the trust center that a new patch was released and begins to investigate the patch on the Technical Support web page.

The ArcGIS Enterprise environment does not have the ability to connect to the internet, so the administrator must download the patch that matches their ArcGIS Enterprise version as well as their operating system. While the patch is downloading, Victor reads the installation instructions on the patch page to check for any special instructions outside of installing the patch. In this instance, all users will need to clear their browser's cache to complete the update.

With the patch downloaded, Victor accesses the ArcGIS Enterprise development environment. To get a general sense of how long the update will take, Victor starts by taking a full backup of ArcGIS Enterprise, including a WebGISDR backup and a system image. After this, Viktor sets ArcGIS Enterprise into read-only mode to simulate environment status.

He begins to install the patch as prescribed in the product documentation. The patch installs successfully as indicated by the installation wizard reporting completion, and Viktor removes the read-only mode from the development environment. He verifies that key applications are accessible and that basic workflows, such as publishing hosted feature layers, are available. The operation took about two hours, factoring in taking backups, installing the patch, and verifying that the system works.

With this information, Viktor begins to inform his user base and other stakeholders of the impending downtime. He starts by approaching his IT team to coordinate an installation time for the patch. It's agreed that conducting the update after business hours will have the least impact on everyone's day-to-day operation. Viktor sets an information banner, which lists the date and time of the downtime, as well as the estimated time needed to install the patch.

As the date approaches, Viktor ensures that an IT representative is on call, ready to help if something unexpected occurs. Because the development and production environments mirror each other, Victor implements the same action plan. He starts by taking a backup of the production environment and setting ArcGIS Enterprise to read-only mode. Victor then installs the security update patch and verifies system availability and workflow stability.

Recalling that the Portal for ArcGIS security patch required an added step after installation, Viktor opts to update the informational banner to include instructions on how to clear users' browser cache. Taking this further, Viktor also tells his IT teammate about the requirements, preparing that department for a possible influx of tickets and questions.

Through Viktor's careful work, SuperBiz International's ArcGIS Enterprise environment was patched to include important security updates. In addition to a successful application, Viktor's work in testing, verifying, and communicating the installation of this patch built trust with the ArcGIS Enterprise user base.

19

Tutorial 19: Install a patch for ArcGIS Enterprise

Create a patch application plan for your ArcGIS Enterprise organization

1. Prepare your organization's downtime rules or expectations in writing.
   - On what days or times will downtime be least disruptive?
   - How long can the system be down before it becomes unacceptably disruptive?
   - What will you do if the system is down longer than acceptable?

2. Create or review a backup strategy for ArcGIS Enterprise.

   > *Tip: See chapter 18 for a full discussion of backup and restore strategies.*

3. Create a communication plan.
   - Who needs to be informed about the impact of the patch?
   - How far in advance do they need to be notified?
   - What level of detail do they need to be provided?
   - How will you distribute the information?
   - Do you need to use multiple distribution channels to ensure the information reaches everyone?

Install available patches

For ArcGIS Enterprise systems that have access to the internet, you can use the patch notification utility to check for and install available patches.

1. Sign in as a user with administrator privileges to the machine where ArcGIS Server is installed.

2. Navigate to the ArcGIS Server installation directory and then move to the **tools/patchnotification** directory.

3. Run the `patchnotification` executable.

4. Click the link for any available patch to read the details of issues addressed by the patch.

5. When you are ready, in accordance with your patch application plan, click the **Install Security Patches** or **Install All Patches** button to apply the patch.

6. From the **Install Patches** window, choose the patches you wish to install and then click **Start**.

Summary

This chapter covered how to prepare and install patches for ArcGIS Enterprise. Taking back-ups, installing patches in lower tier deployments, planning and communicating downtime, and verifying the efficacy of the patch post installation will build your user base's confidence in your ability to maintain a deployment that is secure and up-to-date.

19

CHAPTER 20

Upgrading ArcGIS Enterprise

Objectives

- Use the ArcGIS Enterprise Product Life Cycle to inform upgrade decisions.
- Determine whether your ArcGIS Enterprise deployment meets the upgrade prerequisites.
- Follow the upgrade process.

Introduction

Every new version of ArcGIS Enterprise brings changes to the capabilities of the system. In this chapter, you will learn the concepts and processes you need to upgrade to a new version.

Why upgrade?

The first question you need to answer is whether you should upgrade at all. It is not a good idea to upgrade just because a new version of the software has been released. You want to carefully consider the benefits that a new version would offer, and upgrade if it makes sense for your organization.

New capabilities

The obvious reason to upgrade is that your organization has needs that aren't met by your current version of ArcGIS Enterprise but that would be met by some new capability available with a newer version. For example, an organization may need to audit changes made to the properties of hosted feature services. By upgrading to ArcGIS Enterprise 11.5 or later, the organization can use the built-in audit logging capability of ArcGIS Server to meet this need.

To effectively take advantage of new capabilities, you need to be aware of your organization's evolving business needs. It may be the case that the capabilities of your current version met the needs you had in the past but are no longer adequate to the task because of changes in your needs. Regularly evaluate your business needs and compare them against the capabilities offered by your version of ArcGIS Enterprise.

In addition to your own changing needs, you need to stay current on the changing capabilities of ArcGIS Enterprise. Esri communicates those changes in blog posts, emails, and documentation pages for each version. Review the new capabilities to see whether they provide the functionality that your organization requires but that has not been previously available.

The ArcGIS Enterprise product life cycle

You also need to be aware of the life cycle for a version of ArcGIS Enterprise. As time passes since the original release of a version, it will go through different levels of support:

- General availability means that the version is fully supported.
- Extended support means that new environments for the version are no longer certified.
- Mature support means that software updates and patches are no longer created.
- Retired means that Esri technical support will no longer be able to assist you.

It is best practice to make sure that your version of ArcGIS Enterprise is still receiving updates and patches. For that reason, you want to avoid falling into mature or retired status. Esri publishes the product life cycle for each version, and you should know when your version is scheduled to go into mature support.

Versions of ArcGIS Enterprise alternate between short-term and long-term support releases. A short-term support release will be in general availability for about 18 months, whereas a long-term support release will be in general availability for about two years. Only long-term support releases have extended support, which lasts for roughly another two years. Short-term support releases go directly to mature status after the period for general availability. For that reason, if you need to stay on a single version of ArcGIS Enterprise for an extended period of time, you should choose a long-term support release.

Bug fixes

Even if your version of ArcGIS Enterprise has the GIS capabilities to meet all your business needs, it may have software defects that affect its usability. Later versions may fix those defects, and those fixes might not be available as patches for your current version. In addition to reviewing the new capabilities of each version, check the list of bug fixes to see if a new version would address software defects that could affect your deployment.

These bug fixes can include addressing security vulnerabilities, which makes them particularly important to implement. If the vulnerability is not fixed in a patch for your version, your organization's security policies may require that you upgrade ArcGIS Enterprise to address the vulnerability.

Fictional user story

Medio County has succeeded in expanding the use of geospatial web apps across the organization. In some respects, it has been too successful—a proliferation of publicly accessible apps created with ArcGIS StoryMaps has caused some confusion for county residents who do not know which web pages are authoritative resources.

Elise Medina, the Medio County GIS administrator, has tried addressing the issue by enabling access to ArcGIS StoryMaps only for members of the organization who should be creating these public resources and can be trusted to do so correctly. Unfortunately, the county is using a version of ArcGIS Enterprise in which the granular level of control over app access is not available. The only way to remove access to the app is to also remove access to other functionalities that these members require to do their jobs.

Elise has set up policies and training to help users understand how to publish public apps and keep them updated. But the problem continues to proliferate, and Elise is spending more time than she wants in identifying published apps that should not be publicly accessible.

A newer version of ArcGIS Enterprise does have the granular app permissions that Elise needs. Because of the harm caused by inconsistent or outdated apps, Elise determines that the benefits of upgrading are worth the effort. After the upgrade, Elise is able to ensure only a small set of members have access to the various app builders, but everybody has the other permissions they need. The new capability of the upgraded version isn't a replacement for the policies and training Elise set up, but she can enforce them more easily.

20

Why not upgrade (yet)?

The product life cycle for each version of ArcGIS Enterprise means that you will certainly have to upgrade at some point. But that does not necessarily mean you should upgrade as soon as a new version comes out. There are several good reasons to delay an upgrade:

- **Retired capabilities:** New versions of ArcGIS Enterprise sometimes retire capabilities. For example, version 11.0 removed the ability to publish services from ArcMap. Those retired capabilities might represent critical business workflows that your organization is not yet ready to replace. In that situation, you should not upgrade ArcGIS Enterprise right away. Instead, you should develop a plan to move away from the retired capabilities and then upgrade when your organization is ready.

- **Operational stability:** The operational staff in your organization may require training to make effective use of new capabilities of an upgraded version of ArcGIS Enterprise. It may take them some time to become fully proficient in new workflows or with using new tools. That transition period increases the risk of operational instability because people are more likely to make mistakes and to work more slowly than they did in a familiar system. The more users of ArcGIS Enterprise you have who are affected by changes, the higher that risk is. For that reason, you may want to delay upgrading ArcGIS Enterprise until you have a transition and training plan in place.

- **System downtime:** The most straightforward way to upgrade ArcGIS Enterprise is to run the setup executable for each component on the machine where that component is installed. This has the downside of making your ArcGIS Enterprise deployment unavailable for the duration of the upgrade process. You can deal with this downtime either by scheduling the upgrade for a time when availability is not important or by creating a parallel deployment to handle traffic during the upgrade. Either way, you may want to delay upgrading until you have thoroughly tested your upgrade strategy.

- **Security review:** Some organizations have policies in place that require them to carefully review new software versions for security vulnerabilities. The review itself may be time-consuming. The review may also discover characteristics of the new ArcGIS Enterprise version that are not acceptable in the organization's environment. Either way, the security review may delay upgrading.

Fictional user story

As a company that supplies critical logistics services, MegaBiz has stringent security requirements for all the software it deploys on its systems. Because several of the routing and tracking services published to its ArcGIS Enterprise organization are mission-critical, it also has a strong need for high service uptime. Because thousands of employees use ArcGIS Enterprise services every day, changes to the way the MegaBiz deployment works would require a substantial effort to make everybody aware of all the updated processes. The company's mature web GIS operation also would not benefit significantly from updated capabilities because ArcGIS Enterprise already does everything MegaBiz needs it to do.

For these reasons, Harjeet Singh, the MegaBiz director of data, has decided that the organization is best served by staying on a given version of ArcGIS Enterprise for a relatively long period of time. It upgrades roughly once every three years to the latest long-term support version. That is frequent enough to ensure its software is always in general or extended availability, which meets its security needs to keep receiving patches.

But the GIS administrators on Harjeet's team aren't sitting there doing nothing for three years. They maintain a test environment to help them evaluate every new version of ArcGIS Enterprise. They use that environment to test new capabilities and perform the necessary security audits. They regularly reevaluate MegaBiz's needs and compare those needs to new software capabilities. They have in the past decided to upgrade ahead of schedule because new business needs weren't being met by their existing deployment. They develop process and training improvements so that when the upgrade occurs, MegaBiz employees can quickly be brought up to speed on any changes in their workflows. When a decision has been made to upgrade to a new version, they rigorously test their upgrade process before they use it in their production environment.

Effects on other software from upgrading ArcGIS Enterprise

Your ArcGIS Enterprise deployment probably interacts with other software and systems. Upgrading your ArcGIS Enterprise version can have effects on how, or even whether, it is able to interact with those other systems.

- **Databases:** If you publish services from a user-managed database, be aware of the version compatibility between ArcGIS Enterprise and the Database Management System (DBMS) you use for that database. For example, ArcGIS Enterprise 11.2 supports PostgreSQL version 12, but upgrading to ArcGIS Enterprise 11.3 would require upgrading PostgreSQL to at least version 13. Esri publishes documentation on version compatibility for all supported DBMSs. Make sure you check that documentation before you upgrade ArcGIS Enterprise and update your DBMS if necessary. If your DBMS is used for other systems in your organization, it may be the case that an upgrade has cascading effects on those other systems as well. For example, a retail business's

20

inventory management system may use a database in the same DBMS that holds their enterprise geodatabase. Upgrading the DBMS may also require upgrading the inventory management system. Because the interconnected nature of these systems could be complex, take your time to fully consider the effects of system upgrades.

- **ArcGIS Pro:** If you publish Python script tools from ArcGIS Pro as web tools to ArcGIS Enterprise, you need to be sure that the Python environment is compatible in both. Each version of ArcGIS Pro and ArcGIS Server comes with a specific version of Python, and a particular set of Python packages. If these Python environments are different, the web tool may produce different results from the script tool run locally from ArcGIS Pro. Esri publishes documentation for geoprocessing service compatibility. Make sure you review that documentation before upgrading ArcGIS Enterprise. For users that are publishing web tools, you should help them make sure they are using a version of ArcGIS Pro that is compatible with your version of ArcGIS Enterprise.
- **Custom extensions:** If you have custom server object extensions (SOE) or server object interceptors (SOI), those custom extensions might not be compatible with an upgraded version of ArcGIS Enterprise. Extension developers may need to recompile or even rewrite the project. Sometimes those developer resources are not available, especially when relying on third-party extensions. Work closely with the development teams to avoid the need to make a hard choice between losing the capabilities of the extension by upgrading or missing out on the updated capabilities of ArcGIS Enterprise by not upgrading.

Upgrade prerequisites

Before performing an upgrade, there are several prerequisite steps you should take to ensure a successful upgrade:

- **Review system requirements:** New versions of ArcGIS Enterprise may have new system requirements. Verify that your environment meets any new requirements for supported operating system version, CPU, memory, and disk space for each component you are upgrading. If you need to update your environment, make sure you do that before upgrading ArcGIS Enterprise. Organizations sometimes migrate environments for reasons unrelated to upgrading ArcGIS Enterprise. For example, an organization may decide to move from an on-premises deployment to a public cloud environment. These migrations are an excellent time to consider upgrading ArcGIS Enterprise versions because any new system requirements can be more easily met when setting up a new environment than updating an environment in place.
- **Set up a test environment:** It is possible to encounter a problem during the upgrade process that leaves your deployment in an unusable state. To mitigate that risk, it is best practice to create a separate environment that mirrors your production

environment and then test your upgrade process on that mirrored environment. Testing will help you identify fixes for problems you encounter without the risk of harming your production environment. The practice you get by testing the upgrade will also help you perform the upgrade on your production environment more efficiently.

- **Backup:** You should take a backup of your entire ArcGIS Enterprise deployment. Chapter 18 has details on the considerations for backup and restore processes. Even if you rigorously tested your upgrade process, there is still a nonzero probability that minor differences between your testing and production environments could result in a nonfunctional production system. The less testing you were able to perform, the higher the probability of failure. If the system ends up in a broken state, you will need the backup to restore the previous state of the deployment. Then you can investigate the source of the problem and address it before attempting to upgrade again.

Upgrade order

Esri recommends a particular order for upgrading ArcGIS Enterprise components. Although the software won't necessarily stop you from upgrading in a different order, this is the order in which Esri tests upgrades. If you run into problems using a different order, Esri technical support will probably ask you to start again from the beginning with the recommended order, as follows:

1. Portal for ArcGIS
2. ArcGIS Web Adaptor for the ArcGIS Enterprise portal
3. ArcGIS Server for the hosting server site
4. ArcGIS Web Adaptor for the hosting server site
5. ArcGIS Data Store for the relational data store
6. ArcGIS Data Store for any other ArcGIS-managed data stores
7. All other federated ArcGIS Server sites and their ArcGIS Web Adaptors

Automating upgrades

If you automated the initial deployment of ArcGIS Enterprise using ArcGIS Enterprise Builder, Chef, PowerShell DSC, or one of the cloud builders, you can use that same tool to upgrade ArcGIS Enterprise as well. Automation can reduce human error and system downtime, so it is a good choice if you need to upgrade a complex environment, have high uptime requirements, or need to frequently upgrade environments.

20

Fictional user story

The Becken Pond Conservation Society's deployment of ArcGIS Enterprise was originally set up using Enterprise Builder and for several years was never expanded beyond the original single-machine deployment. Jim Yazzie, the BPCS GIS expert, has always used Enterprise Builder to keep the deployment updated as new versions were released.

Recently, however, Jim migrated the portal web adaptor to a different machine to support a subdomain name change for the ArcGIS Enterprise deployment. One consequence of that decision is that it is no longer possible to use Enterprise Builder to update the deployment. Because of budgetary constraints, BPCS has never had the resources to support a separate test environment, which has always made Jim a bit nervous when doing upgrades. Now that he no longer has an automated solution to rely on, he's even more concerned about the inability to fully test the upgrade before deploying it in production.

Linda Jackson, the BPCS executive director, listens to Jim's concerns, but the budget for a test environment isn't available this year. They are considering delaying the upgrade, but some new capabilities at the most recent version would be helpful for BPCS field scientists. In the end, the deciding factor is that BPCS does not have anything running in ArcGIS Enterprise that will cause a serious problem if it is unexpectedly down for a few days. Jim tells Linda that he can take backups using virtual machine (VM) snapshots and WebGISDR, which should enable him to restore the deployment from backup in case of a failure.

Before the upgrade, Jim takes the production environment down to test the restore process. This test is successful, so Jim and Linda decide to move ahead with the upgrade. The upgrade goes smoothly, but Linda can see that Jim is more nervous than usual while it's under way. She agrees to add a test environment to next year's budget to improve the chances that future upgrades will be successful.

Tutorial 20: Prepare for an upgrade

Upgrading ArcGIS Enterprise is a task for which you must adequately prepare. In this tutorial, you will gather the information you need to enable a successful upgrade.

Review your current deployment

To effectively upgrade your deployment, you need to know about the deployment you already have.

1. Navigate to the Portal Administrator Directory for your deployment and sign in as a user with administrative privileges.

   > *Tip: This directory will generally be located, for example, at https://gis.example/portal/portaladmin.*

2. On the **Site Root** page, determine the version number listed next to **Version**.
   - What version of ArcGIS Enterprise is your current deployment?

3. Navigate to the Enterprise organization page and sign in as a user with administrative privileges.

4. On the navigation bar, click the **Organization** tab. Then click the **Settings** tab.

5. On the left, click the **Servers** section to show the list of federated server sites.
   - Which server roles do you have federated?
   - Do the current federated servers validate? Make sure you address errors before upgrading.

6. For the server site with the **Hosting Server** role, click the **Administration URL** link to open **Server Manager** for that site.

7. At the top of the page, click the **Site** tab and then open the **Data Stores** section.

8. For each user-managed data store registered with the server site, answer the following questions:
   - What type of data store is it?
   - Does it validate?
   - Is this data store still in use?
   - What type and version of DBMS does it use (if any)?

20

Note: ArcGIS Enterprise does not report the answer to the last question. If you do not know the answer, you will need to consult with somebody else in your organization who has the information, or else use the tools of your user-managed data store to find out.

---

9. On the **Site** tab, click the **Software Authorization** tab.
   • Are any federated server sites using a version of ArcGIS Server that is different from the version number used by your ArcGIS Enterprise portal?

10. Navigate to the ArcGIS Enterprise Life Cycle page at links.esri.com/GTKEnterprise-Life Cycle.
    • For your version of ArcGIS Enterprise, what is the date given for the beginning of Mature support?
    • For your version of ArcGIS Enterprise, what is the date given for product retirement?

11. Log on to each of the machines that are running ArcGIS Enterprise components (Portal for ArcGIS, ArcGIS Server, ArcGIS Data Store, ArcGIS Web Adaptors) and answer the following questions:
    • What is the operating system version used by the machine?
    • How many CPU cores does the machine have?
    • How much memory (RAM) does the machine have?
    • How much disk space does the machine have?

Review what's new in the documentation

Every new release, Esri publishes documentation that describes new capabilities and documents any deprecated or retired capabilities. The **What's new** documentation is a good resource to summarize the differences between software versions.

1. Navigate to the ArcGIS Enterprise documentation page at enterprise.arcgis.com and click the **What's new** link to view the documentation for the latest version.

2. Review the new features for the latest version of ArcGIS Enterprise. Make sure you follow any links for more information about what's new in each component of ArcGIS Enterprise you have in your deployment.

3. At the top of the page, click **Other versions** to open the version drop-down list. Review what's new for every version between your current version and the latest version.

- What business needs does your organization have that aren't currently met by your version of ArcGIS Enterprise?

Tip: You will likely need to collaborate with stakeholders throughout your organization to fully answer this question. If you have not done a full business needs analysis in a while, upgrades provide a good opportunity to review your needs.

- Have any capabilities been added to the software that would address business needs your organization has that aren't met by your current version of ArcGIS Enterprise?
- Have any capabilities you rely on been deprecated or retired in more recent versions?

Review system requirements

Because each new software version may have new system requirements, you need to carefully review those requirements before upgrading.

1. On the top of the **What's new** page, click the **Installation and Deployment** tab. Within the introduction paragraph, click the **system requirements** link.

2. If necessary, switch the documentation to the version you are interested in potentially upgrading to.

3. For each component of ArcGIS Enterprise you have in your deployment, click the link to the system requirements page for that component.

4. For each component, verify whether your environment meets the requirements for the following:
   - Operating system version
   - Hardware requirements

5. Return to the **System Requirements** page. On the left, expand the **Supported databases** section.

6. For each user-managed database you use in your current deployment, review the requirements.
   - Will you need to upgrade your DBMS to use it with an upgraded version of ArcGIS Enterprise?
   - Do you have any other systems that might be impacted by an update to your DBMS?

20

7. Navigate to the geoprocessing services compatibility documentation for ArcGIS Pro at links.esri.com/GTKEnterprise-Geoprocessing.

8. Scroll down to the **ArcGIS Pro and ArcGIS Enterprise compatibility** section.
   - Which version of ArcGIS Pro has a compatible geoprocessing environment to the version of ArcGIS Enterprise you are considering upgrading to?
   - If users need to upgrade ArcGIS Pro, how will you communicate that need to them?

Take the next step

If it's appropriate for your circumstances, upgrade your deployment. We strongly recommend you create a backup and test the upgrade in a staging environment before upgrading.

Summary

In this chapter, you learned the process for upgrading ArcGIS Enterprise to a new version. Because each version of ArcGIS Enterprise will eventually go into mature support and later be retired, you will eventually need to upgrade. But you should think carefully about when and why you upgrade and be aware of the prerequisites you need to perform first.

Because it can't be said enough: Make a backup, verify your restore procedures are working and up-to-date, and test your upgrade process before you upgrade your production environment.

Conclusion

Congratulations, you've completed *Getting to Know ArcGIS Enterprise*! By working through the concepts and tutorials of this book, you've learned many fundamental aspects of ArcGIS Enterprise, including implementation, administration, and troubleshooting skills and techniques. You've also learned methods for using ArcGIS Enterprise to work with data from across your organization. We hope that this book will serve as a valuable reference to ArcGIS Enterprise capabilities as you move forward with your career.

Throughout this book, we've identified many additional resources and means that you can use to continue learning about ArcGIS Enterprise. They include the following:

- Product documentation
- Esri Community
- ArcGIS Architecture Center
- Esri Academy

As ArcGIS Enterprise continues to evolve and change, we've found it helpful to understand that it exists as a system to bring people in your organization together, regardless of their GIS skill sets. From managing users and their content to creating web apps and sources of truth for your organization, ArcGIS Enterprise can help you achieve these goals and more.

About Esri Press

Esri Press is an American book publisher and part of Esri, the global leader in GIS software, location intelligence, and mapping. Since 1969, Esri has supported customers with geographic science and geospatial analytics, what we call The Science of Where®. We take a geographic approach to problem-solving, brought to life by modern GIS technology, and are committed to using science and technology to build a sustainable world.

At Esri Press, our mission is to inform, inspire, and teach professionals, students, educators, and the public about GIS by developing print and digital publications. Our goal is to increase the adoption of ArcGIS and to support the vision and brand of Esri. We strive to be the leader in publishing great GIS books, and we are dedicated to improving the work and lives of our global community of users, authors, and colleagues.

Acquisitions
Stacy Krieg
Claudia Naber
Alycia Tornetta
Jenefer Shute

Product Engineering
Craig Carpenter
Maryam Mafuri

Editorial
Carolyn Schatz
Mark Henry
David Oberman

Production
Monica McGregor
Victoria Roberts

Sales & Marketing
Eric Kettunen
Sasha Gallardo
Beth Bauler

Contributors
Christian Harder
Matt Artz

Business
Catherine Ortiz
Jon Carter
Jason Childs

Related titles

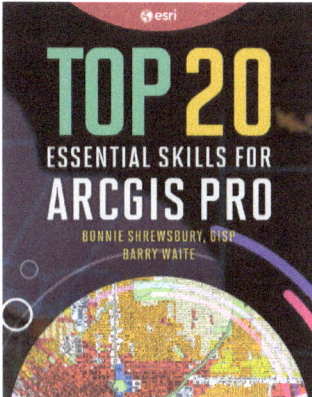

Top 20 Essential Skills for ArcGIS Pro
Bonnie Shrewsbury and Barry Waite
9781589487505

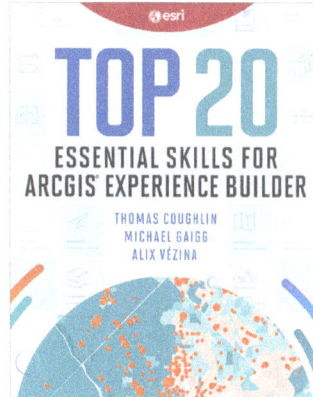

**Top 20 Essential Skills
for ArcGIS Experience Builder**
Thomas Coughlin, Michael Gaigg,
and Alix Vézina
9781589487949

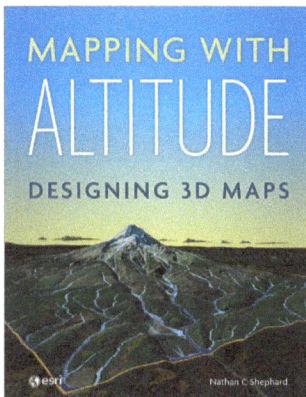

Mapping with Altitude
Nathan C Shephard
9781589485532

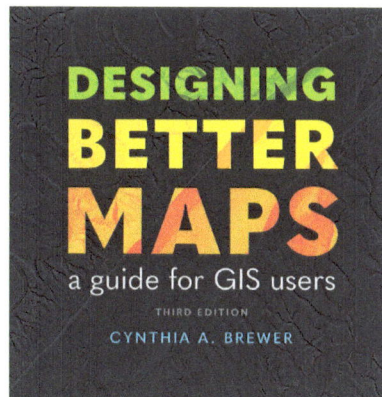

Designing Better Maps
A Guide for GIS Users, third edition
Cynthia A. Brewer
9781589487826

For information on Esri Press books, e-books, and resources, visit our website at
esripress.com